W9-BZY-069

Moonbound

Also by Jonathan Fetter-Vorm

Trinity: A Graphic History of the First Atomic Bomb
Battle Lines: A Graphic History of the Civil War

Moonbound

APOLLO 11 AND THE DREAM OF SPACEFLIGHT

JONATHAN FETTER-VORM

A Novel Graphic from Hill and Wang
A division of Farrar, Straus and Giroux
New York

For Charlotte and Arthur, my fellow travelers of the Earth

Hill and Wang
A division of Farrar, Straus and Giroux
120 Broadway, New York 10271

With additional colors by Laura Martin

Library of Congress Cataloging-in-Publication Data
Fetter-Vorm, Jonathan, 1983– author.
Title: Moonbound : Apollo 11 and the dream of spaceflight / Jonathan Fetter-Vorm.
Description: First edition. | New York : a novel graphic from Hill and Wang, a division of
    Farrar, Straus and Giroux, 2019.
Identifiers: LCCN 2018045286 | ISBN 9780374212452 (hardcover) |
    ISBN 9780374537913 (pbk.)
Subjects: LCSH: Project Apollo (U.S.)—Comic books, strips, etc. | Apollo 11 (Spacecraft)—
    Comic books, strips, etc. | Space flight to the moon—Comic books, strips, etc. |
    Graphic novels.
Classification: LCC TL789.8.U6 A53325 2019 | DDC 629.45/4—dc23
LC record available at https://lccn.loc.gov/2018045286

Our books may be purchased in bulk for promotional, educational, or business
use. Please contact your local bookseller or the Macmillan Corporate and
Premium Sales Department at 1-800-221-7945, extension 5442, or by e-mail at
MacmillanSpecialMarkets@macmillan.com.

www.fsgbooks.com
www.twitter.com/fsgbooks • www.facebook.com/fsgbooks

1 3 5 7 9 10 8 6 4 2

Contents

Foreword

I love books. I started reading them when I was very young, comic books at first. I followed Flash Gordon into the caverns of Mongo, worrying about him every step of the way. Later I added books without any pictures, just full of words. Today I read a lot of those, and I am happy now to add *Moonbound*, which combines the best features of my early and present reading.

In the case of *Moonbound*, I cheated. In addition to just reading this accurate account of flying to the moon, I actually did it, although I did have some help from Neil Armstrong and Buzz Aldrin. Along our way to the moon, we had no time to consider the lives of people like Konstantin Tsiolkovsky, Hermann Oberth, and Robert Goddard, even though we knew they had made our trip possible.

Now you and I can do both simultaneously, reading and seeing the amazing history of spaceflight and how it applied to the various parts of Apollo 11. *Moonbound* blends the two, the past and the present, beautifully. Jonathan Fetter-Vorm makes the history come alive with his vivid descriptions and visual details. You will find out which astronomer had a brass nose, and why.

I know of no other book about Apollo 11 that is more enjoyable.

Michael Collins,
CM Pilot, Apollo 11

Moonbound

YES, STUDYING THE MOON WAS ENOUGH TO ENCOURAGE CURIOUS THOUGHT, FOR THE MOON WAS A PHENOMENON, THE MOON WAS A VOICE WHICH DID NOT SPEAK, A HISTORY WHOSE RECORD ALL REVEALED COULD STILL REVEAL NO ANSWERS: EVERY PROPERTY OF THE MOON PROVED TO CONFUSE A PREVIOUS ASSUMPTION ABOUT ITS PROPERTY. YES, THE MOON WAS A CENTRIFUGE OF THE DREAM, ACCELERATING EVERY NEW IDEA TO INCANDESCENT STATES. ONE TAKES A BREATH WHEN ONE LOOKS AT THE MOON.

--NORMAN MAILER

THE IDEA IS TO GO TO THE MOON, LAND THERE, TAKE OFF AND GET BACK.

--BUZZ ALDRIN

CHAPTER 1
APOLLO 11

IT'S NICE AND QUIET OVER HERE, ISN'T IT?

PLUNGED IN SHADOW, ON THE FAR SIDE OF THE MOON, CUT OFF FROM ALL CONTACT WITH EARTH,

THREE MEN IN A SPACESHIP ABOUT THE SIZE OF A SEMITRUCK CONSIDERED THEIR NEXT MOVES.

IN A MOMENT, THEIR SPACECRAFT WILL SPLIT IN TWO.

IN THE COMMAND MODULE *COLUMBIA*, PILOT *MICHAEL COLLINS* WILL LINGER IN ORBIT...

EAGLE, THIS IS *COLUMBIA*. DO YOU READ?

...WHILE THE CREW OF THE LUNAR MODULE *EAGLE* DESCENDS TO THE MOON.

BUZZ ALDRIN, PILOT

ROGER, LOUD AND CLEAR.

NEIL ARMSTRONG, MISSION COMMANDER

YOU CATS TAKE IT EASY ON THE LUNAR SURFACE.

WE GOT JUST ABOUT A MINUTE TO GO.

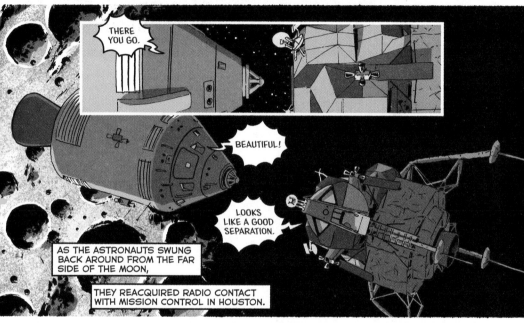

AS THE ASTRONAUTS SWUNG BACK AROUND FROM THE FAR SIDE OF THE MOON,

THEY REACQUIRED RADIO CONTACT WITH MISSION CONTROL IN HOUSTON.

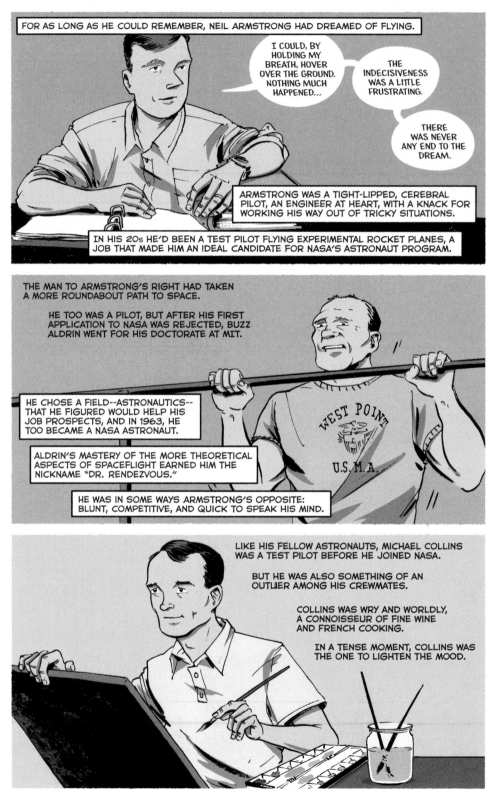

FOR AS LONG AS HE COULD REMEMBER, NEIL ARMSTRONG HAD DREAMED OF FLYING.

I COULD, BY HOLDING MY BREATH, HOVER OVER THE GROUND. NOTHING MUCH HAPPENED...

THE INDECISIVENESS WAS A LITTLE FRUSTRATING.

THERE WAS NEVER ANY END TO THE DREAM.

ARMSTRONG WAS A TIGHT-LIPPED, CEREBRAL PILOT, AN ENGINEER AT HEART, WITH A KNACK FOR WORKING HIS WAY OUT OF TRICKY SITUATIONS.

IN HIS 20s HE'D BEEN A TEST PILOT FLYING EXPERIMENTAL ROCKET PLANES, A JOB THAT MADE HIM AN IDEAL CANDIDATE FOR NASA'S ASTRONAUT PROGRAM.

THE MAN TO ARMSTRONG'S RIGHT HAD TAKEN A MORE ROUNDABOUT PATH TO SPACE.

HE TOO WAS A PILOT, BUT AFTER HIS FIRST APPLICATION TO NASA WAS REJECTED, BUZZ ALDRIN WENT FOR HIS DOCTORATE AT MIT.

HE CHOSE A FIELD--ASTRONAUTICS-- THAT HE FIGURED WOULD HELP HIS JOB PROSPECTS, AND IN 1963, HE TOO BECAME A NASA ASTRONAUT.

ALDRIN'S MASTERY OF THE MORE THEORETICAL ASPECTS OF SPACEFLIGHT EARNED HIM THE NICKNAME "DR. RENDEZVOUS."

HE WAS IN SOME WAYS ARMSTRONG'S OPPOSITE: BLUNT, COMPETITIVE, AND QUICK TO SPEAK HIS MIND.

WEST POINT
U.S.M.A.

LIKE HIS FELLOW ASTRONAUTS, MICHAEL COLLINS WAS A TEST PILOT BEFORE HE JOINED NASA.

BUT HE WAS ALSO SOMETHING OF AN OUTLIER AMONG HIS CREWMATES.

COLLINS WAS WRY AND WORLDLY, A CONNOISSEUR OF FINE WINE AND FRENCH COOKING.

IN A TENSE MOMENT, COLLINS WAS THE ONE TO LIGHTEN THE MOOD.

ALTHOUGH THE PROMISE OF A MAN PLANTING HIS FEET FOR THE FIRST TIME ON ANOTHER WORLD HAD HELPED CONVINCE A NATION TO FUND THIS MISSION TO THE MOON...

...THAT FIRST STEP WAS THE LAST THING ON ARMSTRONG'S MIND AT THE MOMENT.

EAGLE, HOUSTON. IF YOU READ, YOU'RE "GO" FOR POWERED DESCENT.

COPY.

AFTER THE FACT, ARMSTRONG RANKED THE DIFFICULTY OF THE MISSION ON A SCALE OF ONE TO TEN. IF WALKING ON THE MOON WAS A ONE,

THE LUNAR DESCENT WAS PROBABLY A 13.

DURING THE DESCENT, THE EAGLE WOULD BE FALLING DOWN TOWARD THE MOON AT 3800 MILES PER HOUR.

TO SLOW THEIR FALL, THE CREW RELIED ON A ROCKET ENGINE THAT COULD BE THROTTLED UP OR DOWN.

ENGINE ARM: DESCENT. 40 SECONDS.

IS THE CAMERA RUNNING?

CAMERA'S RUNNING.

DESCENT ARMED.

PROCEED... ONE...ZERO...

IGNITION.

IF ALL WENT WELL, THE ONBOARD COMPUTER WOULD CONTROL THE THRUST OF THE ENGINE, WORKING IT LIKE A BRAKE.

AT 30,000 FEET THE CRAFT WAS PITCHED SIDEWAYS TO THE SURFACE OF THE MOON, GIVING THE ASTRONAUTS A CLEAR VIEW OUT THEIR WINDOWS AT WHERE THEY WERE GOING.

IGNITION. 10 PERCENT. THROTTLE UP. LOOKS GOOD!

EAGLE. HOUSTON. EVERYTHING'S LOOKING GOOD HERE. OVER.

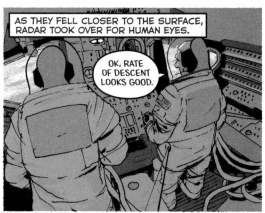

AS THEY FELL CLOSER TO THE SURFACE, RADAR TOOK OVER FOR HUMAN EYES.

OK, RATE OF DESCENT LOOKS GOOD.

FEEDBACK LOOPS BETWEEN THE COMPUTER AND THE RADAR MADE THE DESCENT MOSTLY AUTOMATED.

WHICH WAS GOOD, BECAUSE AS ARMSTRONG HAD LEARNED FROM HIS HOURS IN SIMULATORS, STEERING THIS UNGAINLY MACHINE UNASSISTED WAS DEVILISHLY HARD.

EJECT!

AND UNLIKE IN HIS SIMULATIONS, HERE HE HAD ONLY ONE CHANCE TO GET IT RIGHT.

OUR POSITION CHECKS DOWNRANGE SHOW US TO BE A LITTLE LONG.

ABOUT THREE SECONDS LONG... ROLLING OVER....

AN ALARM MEANT THAT THE COMPUTER WAS STALLING OUT, THAT THE PROCESSOR WAS SOMEHOW OVERWHELMED.

IT COULD BE NOTHING. OR, THIS CLOSE TO THE SURFACE, IT COULD BE CALAMITY.

BACK ON EARTH, MISSION CONTROL SCRAMBLED TO FIGURE OUT WHAT WAS GOING ON.

THE CAREFUL, STEADY VOICE OF GENE KRANZ, THE CONDUCTOR TO THIS ORCHESTRA OF ENGINEERS AND TECHNICIANS, CUT THROUGH THE CONFUSION.

GUIDANCE?

STAND BY...

STEVE BALES, THE CONTROLLER MONITORING THE *EAGLE'S* GUIDANCE SYSTEMS, HAD A CALL TO MAKE.

ABORT THE MISSION OR CARRY ON? OR, IN THE LANGUAGE OF THE CONTROL ROOM, GO OR NO-GO.

HE HAD ONLY SECONDS TO DECIDE.

1202...1202?

THE COMPUTER WAS DESIGNED TO REBOOT IF IT BECAME OVERTAXED.

WHAT'S HAPPENING, WERNHER?

SOMETHING IS WRONG.

SOME SYSTEM ON THE *EAGLE* WAS DEMANDING TOO MUCH FROM THE PROCESSOR.

BALES, LIKE THE OTHER CONTROLLERS, HAD A BRAIN TRUST LISTENING IN FROM A BACK ROOM.

PROGRAM ALARM 1202

AMONG THEM WAS A 25-YEAR-OLD ENGINEER--JACK GARMAN-- WHO'D MEMORIZED ALL THE POSSIBLE ALARM CODES.

ALMOST AS SOON AS HE HEARD THE ALARM, GARMAN HAD AN ANSWER:

1202... IT'S FINE!

AS LONG AS IT DOESN'T REOCCUR, IT'S FINE.

...IT'S FINE.

WE ARE GO ON THAT, FLIGHT.

IF IT DOESN'T REOCCUR, WE'LL BE GO.

EAGLE, WE'RE GO ON THAT ALARM.

AT 7500 FEET, THE *EAGLE* PITCHED UPRIGHT. THEY WERE FALLING AT 100 FEET A SECOND.

DESPITE THE ALARM, THE COMPUTER WAS STILL IN CONTROL. NEVERTHELESS, ARMSTRONG SCANNED THE HORIZON FOR HAZARDS.

BY SIGHTING THROUGH A GRID ON THE WINDOW LIKE A GUNSIGHT, HE COULD ANTICIPATE WHERE THE COMPUTER WAS TAKING THEM.

THE COMPUTER KNEW THEIR ALTITUDE AND TRAJECTORY, BUT IT COULDN'T "SEE" THE RUGGED LANDSCAPE BELOW.

IF ANYTHING WENT WRONG, THEY COULD STILL ABORT THE LANDING, BUT EVEN THAT WAS A RISKY (AND UNTESTED) LAST RESORT.

EAGLE, YOU'RE LOOKING GREAT.

OK. 5000 FEET. 100 FEET PER SECOND.

HERE IT WAS, THE DECISIVE MOMENT. KRANZ POLLED HIS TEAM.

ALL FLIGHT CONTROLLERS, GO/NO-GO FOR LANDING:

RETRO, GO!

FIDO, GO!

GUIDANCE, GO!

CONTROL, GO!

GNC, GO!

EECOM, GO!

SURGEON, GO!

CAPCOM, WE'RE GO FOR LANDING.

3000 FEET. PROGRAM ALARM.

ROGER. WE'RE GO. SAME TYPE. WE'RE GO.

2000 FEET.

ARMSTRONG SAW NOW WHAT HE'D BEEN DREADING.

THE GUIDANCE COMPUTER, IN ITS BLIND CALCULATIONS, HAD STEERED THEM TOWARD THE LIP OF A YAWNING CRATER.

ARMSTRONG REACTED ON INSTINCT.

I'M GOING TO TAKE OVER MANUAL CONTROL.

300 FEET, DOWN 3 1/2 FEET PER SECOND. SLOW IT UP.

HOW'S THE FUEL?

8 PERCENT.

NOT YET RUNNING ON FUMES, BUT CLOSE.

AND NOW THE DEAD MAN'S CURVE, TOO LOW TO ABORT. THEIR CHOICE WAS TO LAND OR CRASH.

GONNA BE RIGHT OVER THAT OTHER CRATER.

I GOT THE SHADOW!

100 FEET. 60 SECONDS OF FUEL LEFT.

40 FEET DOWN. PICKING UP SOME DUST.

DRIFTING TO THE RIGHT A LITTLE.

20 FEET.

FUEL: 30 SECONDS LEFT.

SHUTDOWN.

FROM ORBIT TO TOUCHDOWN, THE LANDING TOOK LESS THAN 10 MINUTES.

OK. ENGINE STOP.

ENGINE ARM IS OFF.

HOUSTON, TRANQUILITY BASE HERE.

THE EAGLE HAS LANDED.

ROGER...TRAN... TRANQUILITY. WE COPY YOU ON THE GROUND.

YOU GOT A BUNCH OF GUYS ABOUT TO TURN BLUE.

WE'RE BREATHING AGAIN.

WE MADE IT.

OK. LET'S GET ON WITH IT.

CHAPTER II
THE MOON

IT HAS AN ADMIXTURE OF COLD AND OF EARTH. IT HAS A
SURFACE IN SOME PLACES LOFTY, IN OTHERS LOW, IN
OTHERS HOLLOW. AND THE DARK IS MIXED TOGETHER
WITH THE FIERY, THE JOINT EFFECT BEING AN IMPRES-
SION OF THE SHADOWY; HENCE IT IS THAT THE MOON IS
SAID TO SHINE WITH A FALSE LIGHT.

--ANAXAGORAS

IT RISES AND IT SETS,

CUTTING A BROAD, CONSTANT ARC ACROSS THE SKY,

SOMETIMES IN DAYLIGHT, SOMETIMES AT DARK,

AND IT HAS MADE THIS JOURNEY FOR 4.5 BILLION YEARS,

ALMOST AS LONG AS THE EARTH ITSELF HAS EXISTED.

THE ORIGINS OF THE MOON ARE A MYSTERY.

THE MOST WIDELY ACCEPTED THEORY GOES LIKE THIS:

BEFORE THE MOON, THERE WAS THEIA,

A PLANET-SIZED OBJECT CAREERING INTO THE EARLY EARTH.

THE REMNANTS OF THAT COLLISION SPUN OFF AND CONGEALED IN EARTH'S ORBIT,

WHERE, EVENTUALLY, THIS MOLTEN BALL OF ROCK COOLED AND HARDENED.

WITHOUT AN ATMOSPHERE OR SIGNIFICANT MOISTURE, THERE WAS NO WIND OR WATER TO SMOOTH THE ROUGH LANDSCAPE.

METEORS BOMBARDED THE MOON'S SURFACE, PITTING IT, PUMMELING IT, SHROUDING IT IN SHEETS OF COSMIC DUST.

SO IT WAS FOR EONS, UNTIL SOMETHING CHANGED ON EARTH.

FOR WHEN THE FIRST HUMANS LOOKED SKYWARD,

WHAT THEY SAW WAS NOT A SPHERE OF ROCK, GLOWING IN THE SUN'S LIGHT,

WAXING AND WANING AS IT ORBITED THE EARTH.

WHAT THEY SAW WAS NOT AN OBJECT AT ALL,

BUT A SYMBOL.

15

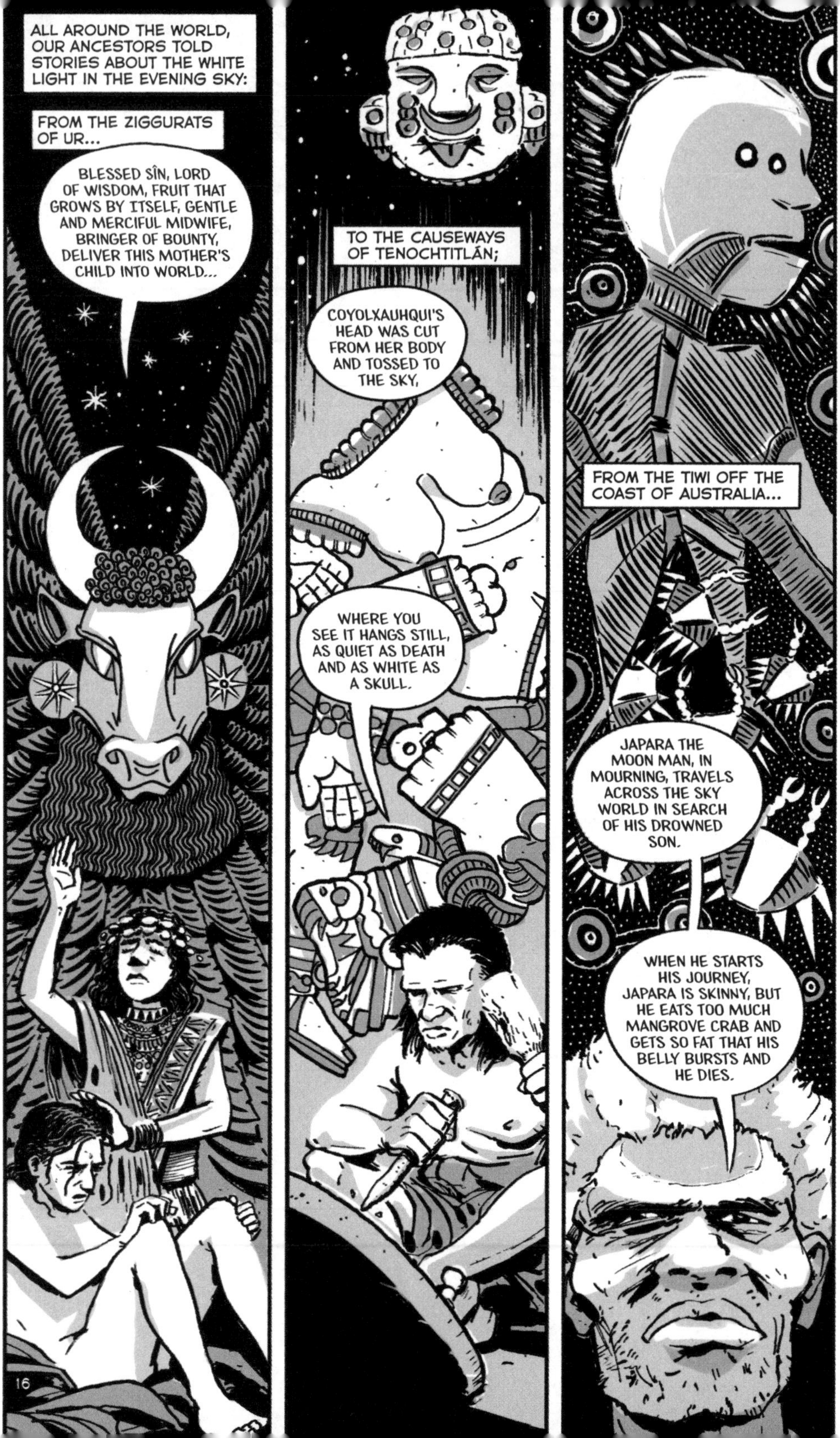
ALL AROUND THE WORLD, OUR ANCESTORS TOLD STORIES ABOUT THE WHITE LIGHT IN THE EVENING SKY:
FROM THE ZIGGURATS OF UR...
BLESSED SÎN, LORD OF WISDOM, FRUIT THAT GROWS BY ITSELF, GENTLE AND MERCIFUL MIDWIFE, BRINGER OF BOUNTY, DELIVER THIS MOTHER'S CHILD INTO WORLD...
TO THE CAUSEWAYS OF TENOCHTITLÁN;
COYOLXAUHQUI'S HEAD WAS CUT FROM HER BODY AND TOSSED TO THE SKY,
WHERE YOU SEE IT HANGS STILL, AS QUIET AS DEATH AND AS WHITE AS A SKULL.
FROM THE TIWI OFF THE COAST OF AUSTRALIA...
JAPARA THE MOON MAN, IN MOURNING, TRAVELS ACROSS THE SKY WORLD IN SEARCH OF HIS DROWNED SON.
WHEN HE STARTS HIS JOURNEY, JAPARA IS SKINNY, BUT HE EATS TOO MUCH MANGROVE CRAB AND GETS SO FAT THAT HIS BELLY BURSTS AND HE DIES.

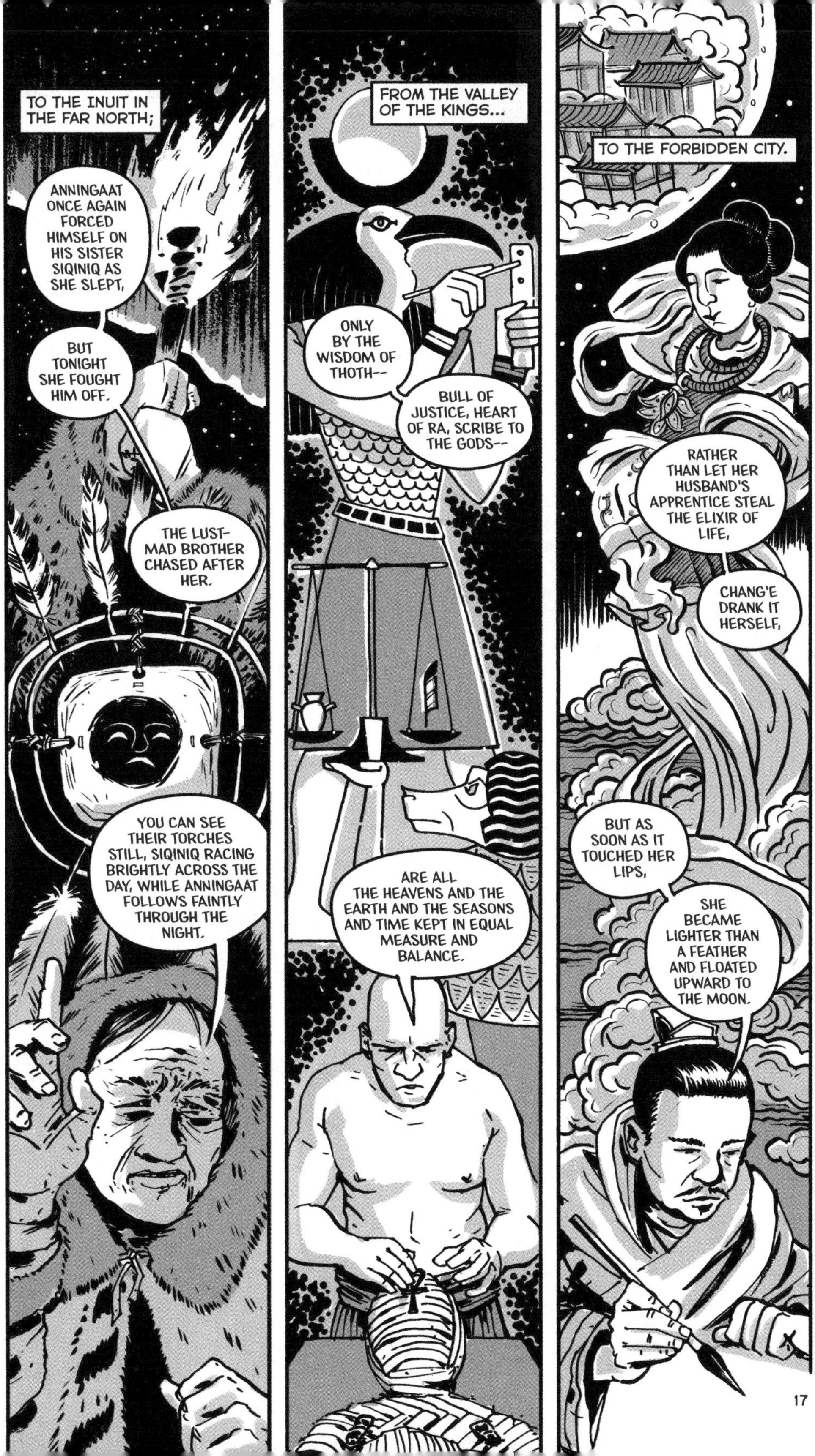
TO THE INUIT IN THE FAR NORTH;
ANNINGAAT ONCE AGAIN FORCED HIMSELF ON HIS SISTER SIQINIQ AS SHE SLEPT,
BUT TONIGHT SHE FOUGHT HIM OFF.
THE LUST-MAD BROTHER CHASED AFTER HER.
YOU CAN SEE THEIR TORCHES STILL, SIQINIQ RACING BRIGHTLY ACROSS THE DAY, WHILE ANNINGAAT FOLLOWS FAINTLY THROUGH THE NIGHT.
FROM THE VALLEY OF THE KINGS...
ONLY BY THE WISDOM OF THOTH--
BULL OF JUSTICE, HEART OF RA, SCRIBE TO THE GODS--
ARE ALL THE HEAVENS AND THE EARTH AND THE SEASONS AND TIME KEPT IN EQUAL MEASURE AND BALANCE.
TO THE FORBIDDEN CITY.
RATHER THAN LET HER HUSBAND'S APPRENTICE STEAL THE ELIXIR OF LIFE,
CHANG'E DRANK IT HERSELF,
BUT AS SOON AS IT TOUCHED HER LIPS,
SHE BECAME LIGHTER THAN A FEATHER AND FLOATED UPWARD TO THE MOON.

THE MOON IS SUCH A POTENT SYMBOL IN PART BECAUSE OF THE WAY IT TRANSFORMS OVER THE COURSE OF A MONTH.

THE CAUSE OF THIS TRANSFORMATION IS SIMPLE: IT'S THE SHADOW CAST BY THE SUN. AS THE MOON ORBITS AROUND THE EARTH, THE ANGLE AT WHICH WE SEE THAT SHADOW CHANGES.

THE MOON CYCLES THROUGH THIS CHANGE EVERY 29½ DAYS.

AND AFTER ABOUT TWELVE CYCLES OF THIS WAXING AND WANING, THE EARTH WILL HAVE ALSO EXPERIENCED EACH OF ITS SEASONS.

IN OTHER WORDS, A YEAR WILL HAVE PASSED.

INDEED, THE CYCLE OF THE MOON'S PHASES WAS PROBABLY HUMANKIND'S EARLIEST WAY OF RECKONING THE PASSAGE OF TIME.

PREHISTORIC BONE FRAGMENTS FROM EUROPE AND AFRICA BEAR HATCH MARKS THAT MIGHT BE THE EARLIEST LUNAR CALENDARS.

EVEN TO AN UNTRAINED EYE, THE MOON CLEARLY HOLDS A SPECIAL PLACE IN THE NIGHT SKY.

BECAUSE IT'S CLOSER THAN ANYTHING ELSE, ITS CHANGES ARE VISIBLE AT A GLANCE.

THE MOON IS BORN EVERY MONTH, INCREASES, IS PERFECTED, DIMINISHES, IS CONSUMED, IS RENEWED.

AS IN THE MOON EVERY MONTH, SO IN RESURRECTION ONCE FOR ALL TIME.

AND BECAUSE IT GLOWS ONLY WITH REFLECTED LIGHT, AND NOT WITH THE FULL GLARE OF THE SUN, WE CAN LOOK AT THE MOON FOR LONG ENOUGH TO WONDER WHAT IT IS WE'RE LOOKING AT.

WHAT, FOR EXAMPLE, ARE THOSE MARKINGS ON THE FACE OF THE MOON?

DO THEY SHOW, AS IN LEGENDS FROM EAST ASIA, A RABBIT WORKING ITS MORTAR AND PESTLE?

OR, AS IN FABLES FROM NORTHERN EUROPE, AN OLD MAN STOOPED BENEATH A BUNDLE OF STICKS?

EARLY MODERN ASTRONOMERS ASSUMED THE DARK SPOTS WERE EVIDENCE OF WATER.

AND ON MOON MAPS THEY LABELED THESE SPOTS *MARIA*, THE LATIN WORD FOR "SEAS."

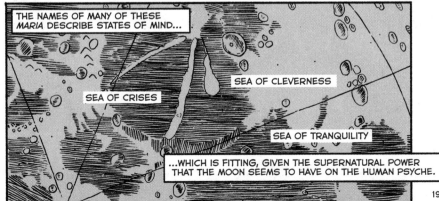

THE NAMES OF MANY OF THESE *MARIA* DESCRIBE STATES OF MIND...

SEA OF CLEVERNESS

SEA OF CRISES

SEA OF TRANQUILITY

...WHICH IS FITTING, GIVEN THE SUPERNATURAL POWER THAT THE MOON SEEMS TO HAVE ON THE HUMAN PSYCHE.

19

THE POWER TO MAKE US FALL IN LOVE...

...TO STIR THE MONSTERS INSIDE US...

...OR TO DRIVE US INTO LUNACY.

THAT THE MOON EXERTS A POWERFUL FORCE ON THE WORLD IS NOT JUST FOLKLORE.

CONSIDER THE RISE AND FALL OF THE TIDES.

THE GRAVITATIONAL PULL OF THE MOON DISTORTS THE WATER IN EARTH'S OCEANS,

CREATING A CEASELESS CYCLE OF BULGES AND DEPRESSIONS: HIGH AND LOW TIDE.

OTHER LUNAR EFFECTS, HOWEVER, ARE HARDER TO PROVE...

IT'S PROBABLY JUST A COINCIDENCE THAT A WOMAN'S MENSTRUAL CYCLE IS ABOUT THE SAME DURATION AS A LUNAR MONTH.

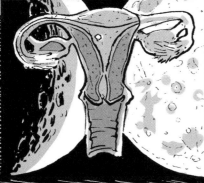

AND WHILE MANY INSECTS, BIRDS, AND AQUATIC ANIMALS ALTER THEIR BEHAVIOR DURING CERTAIN PHASES OF THE MOON,

THOSE CHANGES HAVE MORE TO DO WITH MOONLIGHT AND THE TIDES THAN WITH SOME MYSTERIOUS LUNAR FORCE.

THERE IS ONE WAY, HOWEVER, THAT THE MOON EVOKES AN UNDENIABLY PROFOUND EFFECT: A SOLAR ECLIPSE.

EVERY FEW YEARS, SOMEWHERE ON THE EARTH, THE MOON BRIEFLY BLOTS OUT THE LIGHT OF THE SUN.

THE EARLIEST RECORD OF A SOLAR ECLIPSE COMES FROM THE EMPIRE OF ASSYRIA, IN MODERN-DAY IRAQ.

AMONG THE PEOPLE WHO CULTIVATED THIS FERTILE LAND BETWEEN THE TIGRIS AND EUPHRATES RIVERS WERE ASTROLOGER-PRIESTS WHO WATCHED THE SKY EVERY NIGHT...

...AND RECORDED WHAT THEY SAW IN "ASTRONOMICAL DIARIES."

IN THE YEAR OF BUR-SAGALE OF GUZANA, IN THE MONTH OF SIWAN, THERE WAS AN ECLIPSE OF THE SUN.

FROM THESE TABLETS, WITH RECORDS SPANNING CENTURIES,

THE ASTRONOMERS LEARNED TO PREDICT WHEN AND WHERE THE MOON WOULD APPEAR,

AS WELL AS CERTAIN ECLIPSES.

FOR ANYONE NOT TRAINED IN THIS NEW ART, THE POWER OF ASTRONOMY WAS A TERRIFYING, ALMOST SUPERNATURAL ABILITY.

THE BABYLONIAN ASTRONOMERS FOUND A WAY TO TRANSFORM WHAT HAD ALWAYS SEEMED LIKE RANDOM PHENOMENA INTO PREDICTABLE, AUSPICIOUS EVENTS.

KINGDOMS ROSE AND COLLAPSED ACCORDING TO THE OMENS INSCRIBED IN THE NIGHT SKY.

OMENS THAT ALL THE WORLD COULD SEE, BUT ONLY A SELECT FEW COULD UNDERSTAND.

NOWADAYS, AN ECLIPSE IS MORE OF A NOVELTY THAN A HARBINGER,

THOUGH THE MOON IS STILL A FREQUENT REMINDER THAT
WHAT WE KNOW AND WHAT WE SEE AREN'T ALWAYS THE SAME.

CONSIDER, FOR EXAMPLE, ITS COLOR.

THERE ARE BLOOD MOONS AND BLUE MOONS,

BUT TO OUR EYES THE MOON IS
MOST OFTEN A PEARLY WHITE.

YET THE DUST THAT
POWDERED NEIL
ARMSTRONG'S SPACE
SUIT WAS A GREASY
BLACK, THE COLOR
OF GRAPHITE.

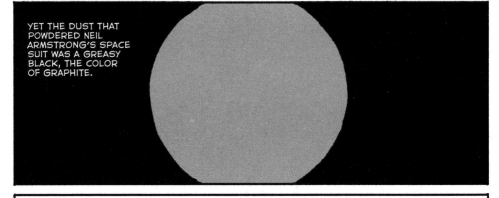

TAKE AWAY THE SUNLIGHT THAT GLINTS
OFF ITS SURFACE AND REMOVE IT
FROM THE BLACKNESS OF DEEP
SPACE...

...AND THE WHITE MOON
LOOKS MORE LIKE A
DULL, DARK ROCK.

THAT THE MOON CAN BE BOTH AT ONCE
IS WHAT MAKES IT SO REMARKABLE.

THE MOON IS MADE OF EQUAL PARTS MINERAL AND MYTH.

IT CAN BE AN OMEN TO FORTUNE-TELLERS AND A MUSE TO POETS,

A THING TO WORSHIP, TO WISH UPON, AND TO MEASURE.

A PALE LIGHT IN THE COLD DARK.

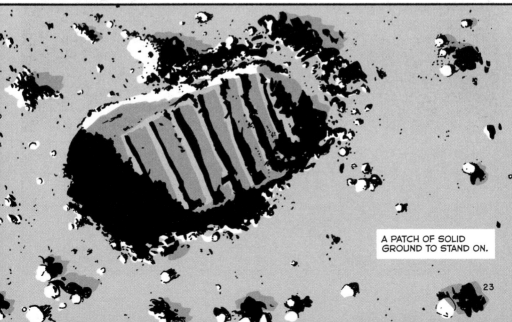

A PATCH OF SOLID GROUND TO STAND ON.

CHAPTER III
EAGLE

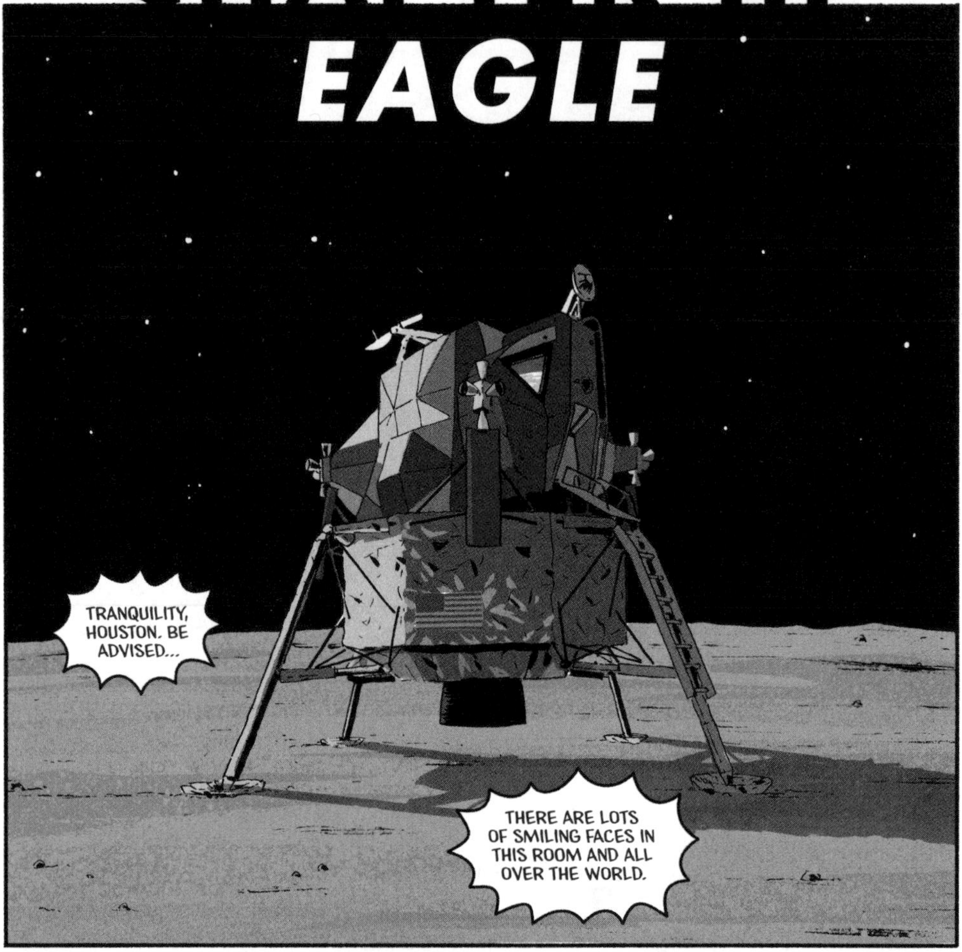
TRANQUILITY, HOUSTON. BE ADVISED...
THERE ARE LOTS OF SMILING FACES IN THIS ROOM AND ALL OVER THE WORLD.

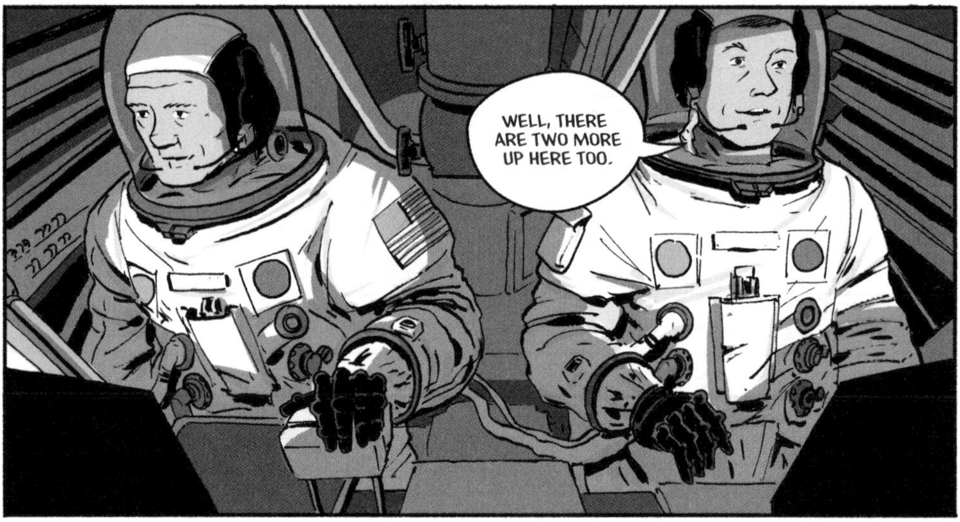
WELL, THERE ARE TWO MORE UP HERE TOO.

AND DON'T FORGET ONE IN THE COMMAND MODULE...

...THANKS FOR PUTTING ME ON RELAY, HOUSTON.

I WAS MISSING ALL THE ACTION.

IT SURE SOUNDED GREAT FROM UP HERE.

YOU GUYS DID A FANTASTIC JOB.

THANK YOU. JUST KEEP THAT ORBITING BASE READY FOR US UP THERE NOW.

HOUSTON, THE GUYS WHO SAID THAT WE WOULDN'T BE ABLE TO TELL PRECISELY WHERE WE ARE ARE THE WINNERS TODAY.

WE WERE A LITTLE BUSY WORRYING ABOUT PROGRAM ALARMS...

ROGER, TRANQUILITY. NO SWEAT. WE'LL FIGURE IT OUT. OVER.

THE FIRST JOB FOR THE FIRST MEN ON THE MOON WAS TO GET READY TO LEAVE.

FOR HALF AN HOUR, THEY PERFORMED A SIMULATED COUNTDOWN, REHEARSING FOR AN EMERGENCY GETAWAY.

DURING LULLS IN THE PROCEDURE, ARMSTRONG AND ALDRIN STOLE QUICK GLIMPSES OF THEIR SURROUNDINGS.

OUT OF THE LEFT-HAND WINDOW IS A RELATIVELY LEVEL PLAIN, CRATERED... WITH SOME SMALL RIDGES.

THERE'S A HILL IN VIEW. DIFFICULT TO ESTIMATE BUT MIGHT BE A HALF A MILE OR A MILE.

IT LOOKS LIKE A COLLECTION OF JUST ABOUT EVERY VARIETY OF ROCK YOU COULD FIND.

THE COLOR IS... WELL, IT VARIES PRETTY MUCH DEPENDING ON THE ANGLE OF THE SUN.

IT'S PRETTY MUCH WITHOUT COLOR. IT'S GRAY, A VERY WHITE, CHALKY GRAY.

CLICK

ALTHOUGH, AS THE ASTRONAUTS WOULD SOON DISCOVER, THE MOON HAD A WAY OF PLAYING TRICKS ON THE EYE.

WE CAN'T SEE ANY STARS OUT THE WINDOW. BUT IN MY OVERHEAD HATCH, I'M LOOKING AT THE EARTH.

IT'S BIG AND BRIGHT AND BEAUTIFUL.

BUZZ IS GOING TO GIVE A TRY AT SEEING SOME STARS THROUGH THE SEXTANT.

LIKE AN ANCIENT NAVIGATOR,

ALDRIN LOOKED TO THE STARS FOR SOME SIGN OF HIS BEARINGS.

NO ONE COULD SAY FOR SURE WHERE THE ASTRONAUTS HAD LANDED.

FOR NOW, THEY WERE ADRIFT,

SOMEWHERE ON THE SEA OF TRANQUILITY.

CHAPTER IV
ASTRONOMERS

MAN HATH WEAVED OUT A NET, AND THIS NET THROWNE
UPON THE HEAVENS, AND NOW THEY ARE HIS OWN...

--JOHN DONNE

MANY THINGS HAPPENED TO JOHANNES KEPLER WHEN HE WAS A BOY, AND MOST OF THEM WERE BAD.

JOH?

SMALLPOX, DOUBLE VISION, SKIN SORES, MANGE, CONSTANT HEADACHES.

HIS FATHER WAS OFTEN ANGRY AND SELDOM HOME.

AND THERE WERE PERSISTENT RUMORS THAT HIS MOTHER WAS A WITCH.

ARE YOU AWAKE? COME.

ONE TIME, HE EVEN FOUND A WORM EMBEDDED IN HIS HAND.

COME. YOU NEED TO SEE THIS.

BUT ON A WARM SUMMER NIGHT IN 1580, WHEN KEPLER WAS NINE, SOMETHING HAPPENED THAT WASN'T BAD AT ALL.

IN FACT, WHAT HAPPENED THAT NIGHT WAS MARVELOUS.

JOHANNES KEPLER SAW HIS FIRST LUNAR ECLIPSE.

KEPLER WAS BORN IN 1571, IN THE CITY OF WEIL DER STADT.

HE SHOWED AN EARLY APTITUDE FOR MATH...

...MUCH TO THE ANNOYANCE OF HIS PARENTS.

THIS BOY NEEDS TO LEARN THE VALUE OF REAL WORK.

COUGH

RELIEF FROM A LIFE OF HARD LABOR CAME WHEN KEPLER TURNED 13 AND ENTERED A SEMINARY.

HE WAS STILL MOSTLY MISERABLE--HE HAD FEW FRIENDS AND MANY BULLIES--BUT AT LEAST NOW HE COULD READ AND WRITE.

HIS CURIOSITY WAS VORACIOUS:

HERE'S AN ESSAY I WROTE ABOUT THE HEAVENS.

AND HERE'S ONE I WROTE ABOUT SPIRITS AND THE ELEMENTS.

AND HERE'S ONE ABOUT THE NATURE OF FIRE. AND THE TIDES.

AND THE SHAPE OF THE CONTINENTS.

THE BREADTH OF HIS INTERESTS WAS ONLY MATCHED BY THE DEPTH OF HIS SELF-DOUBT.

"INCONSISTENCY, THOUGHTLESSNESS, LACK OF DISCIPLINE"

"SUDDEN ENTHUSIASMS WHICH DO NOT LAST"

"HE IS A BITTER HATER OF WORK"

"FAILURE TO FINISH THINGS HE HAS BEGUN"

OUT OF THIS CAULDRON OF FASCINATION AND DOUBT, A PECULIAR NOTION BEGAN TO COHERE.

SOME VARIATION OF IT WOULD LINGER IN THE PERIPHERY OF KEPLER'S IMAGINATION FOR THE REST OF HIS LIFE.

HE CAME TO CALL IT *THE DREAM.*

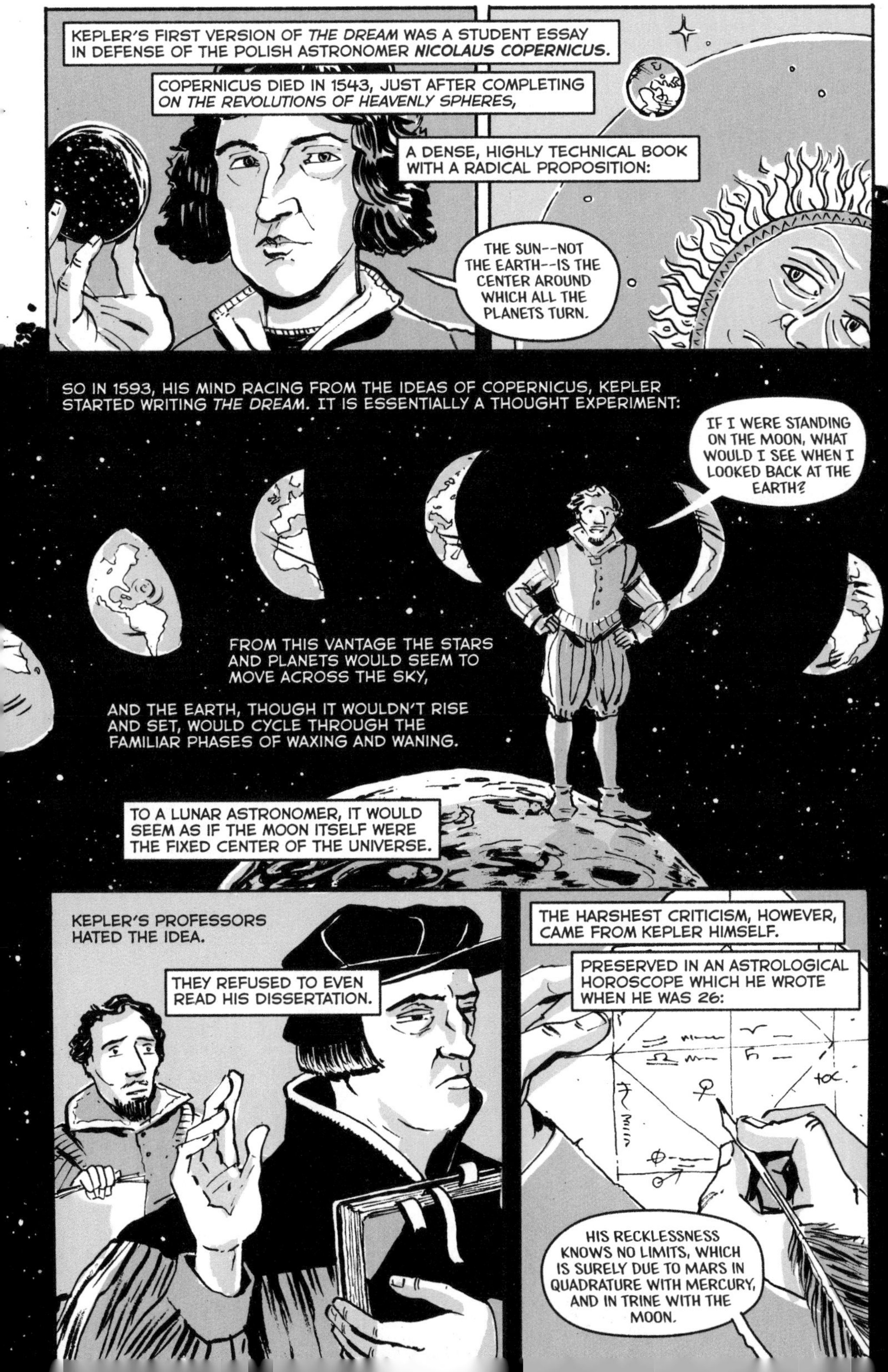
KEPLER'S FIRST VERSION OF *THE DREAM* WAS A STUDENT ESSAY IN DEFENSE OF THE POLISH ASTRONOMER ***NICOLAUS COPERNICUS***.
COPERNICUS DIED IN 1543, JUST AFTER COMPLETING *ON THE REVOLUTIONS OF HEAVENLY SPHERES*,
A DENSE, HIGHLY TECHNICAL BOOK WITH A RADICAL PROPOSITION:
THE SUN--NOT THE EARTH--IS THE CENTER AROUND WHICH ALL THE PLANETS TURN.
SO IN 1593, HIS MIND RACING FROM THE IDEAS OF COPERNICUS, KEPLER STARTED WRITING *THE DREAM*. IT IS ESSENTIALLY A THOUGHT EXPERIMENT:
IF I WERE STANDING ON THE MOON, WHAT WOULD I SEE WHEN I LOOKED BACK AT THE EARTH?
FROM THIS VANTAGE THE STARS AND PLANETS WOULD SEEM TO MOVE ACROSS THE SKY,
AND THE EARTH, THOUGH IT WOULDN'T RISE AND SET, WOULD CYCLE THROUGH THE FAMILIAR PHASES OF WAXING AND WANING.
TO A LUNAR ASTRONOMER, IT WOULD SEEM AS IF THE MOON ITSELF WERE THE FIXED CENTER OF THE UNIVERSE.
KEPLER'S PROFESSORS HATED THE IDEA.
THEY REFUSED TO EVEN READ HIS DISSERTATION.
THE HARSHEST CRITICISM, HOWEVER, CAME FROM KEPLER HIMSELF.
PRESERVED IN AN ASTROLOGICAL HOROSCOPE WHICH HE WROTE WHEN HE WAS 26:
HIS RECKLESSNESS KNOWS NO LIMITS, WHICH IS SURELY DUE TO MARS IN QUADRATURE WITH MERCURY, AND IN TRINE WITH THE MOON.

LATER IN LIFE, KEPLER EARNED A STEADY INCOME AS AN ASTROLOGER TO RICH AND POWERFUL MEN.

HE SECURED EARLY FAME BY PROPHESYING AN UNSEASONABLE FROST.

BUT KEPLER WAS AMBIVALENT ABOUT ASTROLOGY.

TINK

TRUE FORTUNE-TELLING WAS THE DOMAIN OF CHARLATANS AND FALSE PROPHETS,

NOT OF PIOUS LUTHERANS AND PROFESSORS OF MATHEMATICS.

AND YET, IF KEPLER HADN'T BEEN OBSESSED WITH DECIPHERING THE ARCANE MYSTERIES OF THE COSMOS,

HE NEVER WOULD HAVE STUMBLED UPON HIS GREAT INSIGHTS ABOUT THE MOVEMENTS OF PLANETS.

INSIGHTS THAT BECAME THE BEDROCK OF MODERN ASTRONOMY.

KEPLER LIVED AT A MOMENT OF GREAT INTELLECTUAL UPHEAVAL IN EUROPE.

THE DIVINE MYSTICISM OF THE MIDDLE AGES WAS GIVING WAY TO A NEW KIND OF UNDERSTANDING, ONE BASED NOT ON SCRIPTURE BUT ON HUMAN OBSERVATION AND REASONING.

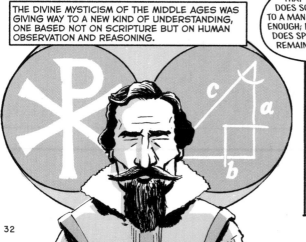

THAT THE SKY DOES SOMETHING TO A MAN IS OBVIOUS ENOUGH; BUT WHAT IT DOES SPECIFICALLY REMAINS HIDDEN.

KEPLER TRIED TO EMBRACE THIS NEW EMPIRICISM WITHOUT GIVING UP ON HIS BELIEF IN THE SACRED ORDER OF THE UNIVERSE.

HE PUBLISHED HIS FIRST BOOK--*THE MYSTERIES OF THE COSMOS*-- WHEN HE WAS 25 AND A PROFESSOR OF MATH.

IT WAS A SPIRITED DEFENSE OF THE COPERNICAN SYSTEM,

THE FIRST TIME, REALLY, THAT A PROFESSIONAL ASTRONOMER SHOWED PUBLIC SUPPORT FOR THE SUN-CENTERED UNIVERSE.

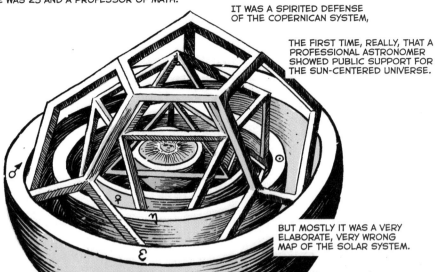

BUT MOSTLY IT WAS A VERY ELABORATE, VERY WRONG MAP OF THE SOLAR SYSTEM.

BURIED WITHIN ITS PAGES, HOWEVER, WAS A TRULY MODERN IDEA: THAT ANY ATTEMPT TO EXPLAIN THE NATURE OF THE PLANETS MUST BE BASED ON OBSERVATION.

IF THESE DO NOT CONFIRM THE THESIS, THEN ALL OUR PREVIOUS EFFORTS HAVE DOUBTLESS BEEN IN VAIN.

AS IT TURNED OUT, KEPLER'S OWN OBSERVATIONS SHOWED THAT HIS SYSTEM WAS A MESS.

TO PUT IT IN MODERN TERMS: THE DATA DIDN'T SUPPORT HIS HYPOTHESIS.

KEPLER WAS NOT PREPARED TO GIVE UP.

BUT HE ALSO RECOGNIZED THAT IF HE WAS GOING TO MAKE AN ACCURATE MODEL,

HE NEEDED MORE INFORMATION ABOUT THE ACTUAL PLANETS.

AND FOR THAT, KEPLER KNEW EXACTLY WHERE TO LOOK.

33

TYCHO BRAHE WAS A DANISH ASTRONOMER, FAMOUS AMONG THE COURTS AND COLLEGES OF EUROPE FOR HIS EXACTING OBSERVATIONS OF CELESTIAL MOVEMENTS.

TYCHO WAS A NOBLEMAN,

WITH A TROVE OF PRICELESS INSTRUMENTS,

A BRASS NOSE (HE LOST HIS REAL ONE IN A DUEL),

AND A LIFETIME'S WORTH OF ASTRONOMICAL DATA,

WHICH HE SHARED WITH PRECISELY NO ONE.

IN 1597, TYCHO FELL OUT OF FAVOR WITH THE NEW KING OF DENMARK AND WAS EXILED FROM HIS HOMELAND.

TYCHO FOUND REFUGE IN THE HOLY ROMAN EMPIRE OF RUDOLPH II IN PRAGUE, A PATRON OF ALL THINGS SCIENTIFIC AND ALCHEMICAL.

HE APPOINTED TYCHO TO THE POSITION OF IMPERIAL ASTRONOMER.

TYCHO'S FIRST ORDER OF BUSINESS WAS TO TRANSFORM AN OLD CASTLE OUTSIDE OF TOWN INTO A STATE-OF-THE-ART ASTRONOMICAL OBSERVATORY.

IN FEBRUARY 1600, KEPLER PAID TYCHO A VISIT. HE RECORDED HIS MOTIVES IN A LETTER:

"TYCHO POSSESSES THE BEST OBSERVATIONS,

"AND THUS, SO TO SPEAK, THE MATERIAL FOR THE BUILDING OF THE NEW EDIFICE...

"...HE LACKS ONLY THE ARCHITECT WHO WOULD PUT ALL THIS TO USE ACCORDING TO HIS OWN DESIGN."

KEPLER JOINED TYCHO'S OBSERVATORY, WHERE HE LEARNED FROM THE MASTER HOW TO MAKE CAREFUL, PRECISE OBSERVATIONS OF THE HEAVENS.

TYCHO PUT HIM TO WORK, BUT REFUSED TO SHARE HIS DATA WITH KEPLER.

THE TWO ASTRONOMERS MADE FOR FIERY COLLEAGUES, BUT ULTIMATELY, THEY NEEDED EACH OTHER.

HIS YOUTHFUL ENERGY IS JUST WHAT I NEED TO COMPLETE MY MAGNUM OPUS.

HIS LIFETIME OF OBSERVATIONS ARE JUST WHAT I NEED TO COMPLETE MY MAGNUM OPUS.

AND THEN, SUDDENLY, IN 1601, TYCHO GREW VIOLENTLY ILL.

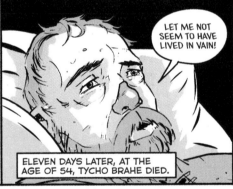

LET ME NOT SEEM TO HAVE LIVED IN VAIN!

ELEVEN DAYS LATER, AT THE AGE OF 54, TYCHO BRAHE DIED.

TWO DAYS AFTER THAT, KEPLER WAS APPOINTED IMPERIAL ASTRONOMER.

AND HE FINALLY ACQUIRED THE PRIZE HE'D BEEN AFTER.

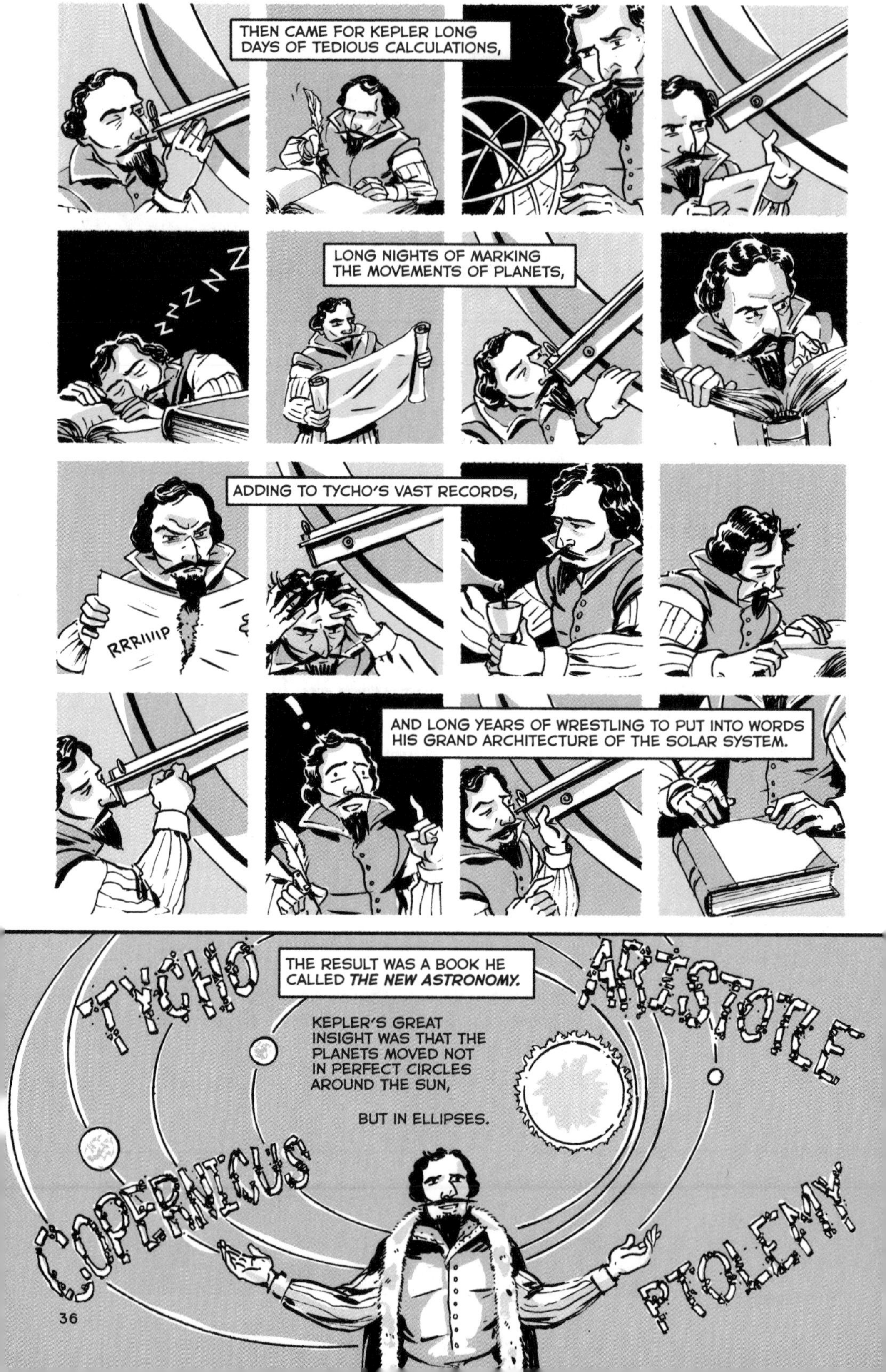
THEN CAME FOR KEPLER LONG DAYS OF TEDIOUS CALCULATIONS,
ZZZZZ
LONG NIGHTS OF MARKING THE MOVEMENTS OF PLANETS,
ADDING TO TYCHO'S VAST RECORDS,
RRRIIIIP
AND LONG YEARS OF WRESTLING TO PUT INTO WORDS HIS GRAND ARCHITECTURE OF THE SOLAR SYSTEM.
THE RESULT WAS A BOOK HE CALLED *THE NEW ASTRONOMY.*
TYCHO
ARISTOTLE
KEPLER'S GREAT INSIGHT WAS THAT THE PLANETS MOVED NOT IN PERFECT CIRCLES AROUND THE SUN,
BUT IN ELLIPSES.
COPERNICUS
PTOLEMY

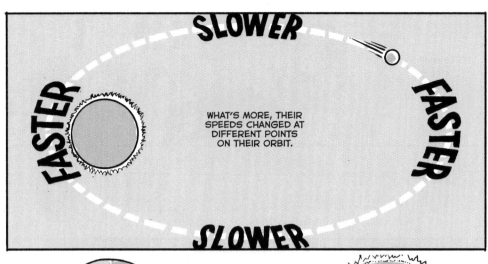

SLOWER

FASTER

FASTER

SLOWER

WHAT'S MORE, THEIR SPEEDS CHANGED AT DIFFERENT POINTS ON THEIR ORBIT.

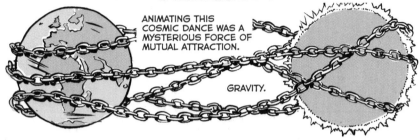

ANIMATING THIS COSMIC DANCE WAS A MYSTERIOUS FORCE OF MUTUAL ATTRACTION.

GRAVITY.

THROUGH A FEAT OF OTHERWORLDLY DILIGENCE,

AND WITHOUT THE BENEFIT OF MODERN MATHEMATICAL TOOLS,

KEPLER HAD DECIPHERED THE FUNDAMENTAL LAWS OF CELESTIAL MOTION,

LAWS THAT WOULD EVENTUALLY ENABLE HUMANS TO LEAVE THE PLANET EARTH.

MEANWHILE, IN THE DUTCH REPUBLIC, 1608:

A GERMAN SPECTACLE MAKER NAMED HANS LIPPERSHEY UNVEILED BEFORE AN AUDIENCE OF ARISTOCRATS A NEW INVENTION.

IT WAS A SIMPLE DEVICE WITH AN UNCANNY POWER.

THE POWER TO MAKE FARAWAY THINGS LOOK IMPOSSIBLY CLOSE.

ONE OF THE MANY NAMES FOR THIS NEW INVENTION WAS *IL CANNONE OCCHIALE*--

THE EYE CANNON.

MY GOD...

MILITARY MEN SAW IN THIS SPYGLASS A BRAVE NEW ADVANCE IN WARFARE.

FROM NOW ON I COULD NO LONGER BE SAFE,

FOR YOU WILL SEE ME FROM AFAR.

THEN WE SHALL SIMPLY FORBID OUR MEN TO SHOOT AT YOU.

NO ONE YET IMAGINED THAT IT WOULD REALIGN OUR UNDERSTANDING OF THE UNIVERSE ITSELF.

MASTER LIPPERSHEY, YOU WILL RECEIVE A HANDSOME SUM FOR THESE GLASSES,

AND FOR ANY OTHERS THAT YOU CAN PRODUCE.

BUT I COMMAND YOU: DO NOT SHARE THIS ART WITH ANYONE.

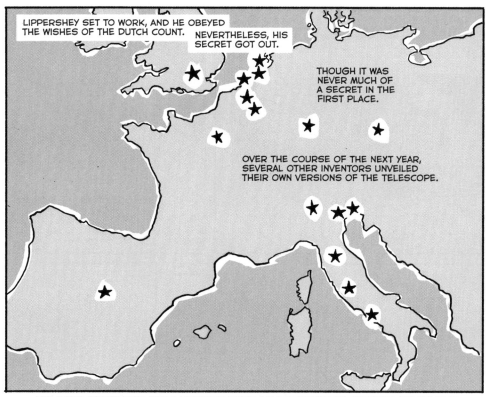

THE SPYGLASS PASSED FROM COURT TO KINGDOM, REACHING THE REPUBLIC OF VENICE IN THE SUMMER OF 1609, PIQUING THE INTEREST OF A NEARBY UNIVERSITY PROFESSOR.

ONE EVENING IN THE FALL OF 1609, GALILEO AIMED HIS NEW TELESCOPE AT THE MOON.

OVER THE COURSE OF SEVERAL NIGHTS OF OBSERVATION, HE DRAFTED A SERIES OF SKETCHES,

REVEALING DETAILS THAT WERE INVISIBLE TO THE NAKED EYE.

DETAILS THAT WEREN'T SUPPOSED TO BE THERE.

ARISTOTLE--

THE TOUCHSTONE OF ASTRONOMY AT THE TIME--

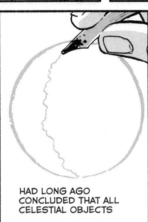

HAD LONG AGO CONCLUDED THAT ALL CELESTIAL OBJECTS

WERE PURE, PERFECT, AND IMMUTABLE,

NOT LIKE THE MESSY MATERIAL OF EARTH.

BUT GALILEO'S CLOSE-UP VIEW OF THE MOON DIDN'T LOOK PURE OR PERFECT AT ALL.

THE MOON IS BY NO MEANS ENDOWED WITH A SMOOTH AND POLISHED SURFACE, BUT IS ROUGH AND UNEVEN.

AND IT IS LIKE THE FACE OF THE EARTH ITSELF, WHICH IS MARKED HERE AND THERE WITH CHAINS OF MOUNTAINS AND DEPTHS OF VALLEYS.

GALILEO ANNOUNCED HIS OBSERVATIONS IN 1610 IN A PAMPHLET TITLED *SIDERIUS NUNCIUS*, OR *THE STARRY MESSENGER.*

IT SOLD OUT IMMEDIATELY.

LIKE THE TELESCOPE ITSELF, *THE STARRY MESSENGER* SPREAD ACROSS EUROPE.

GALILEO WROTE IT IN SIMPLE, CLEAR LANGUAGE...

...YOU DIDN'T HAVE TO BE A MATHEMATICIAN OR ASTRONOMER TO UNDERSTAND IT.

ACTUALLY, YOU HARDLY EVEN NEEDED TO *READ* THE BOOK TO UNDERSTAND IT:

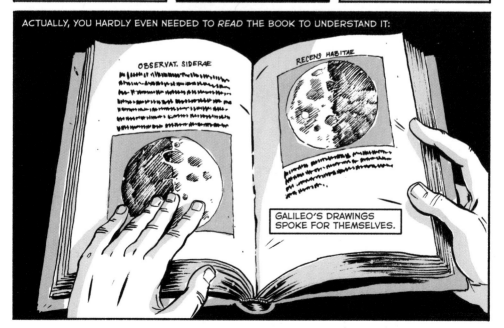

OBSERVAT. SIDERAE

RECENS HABITAE

GALILEO'S DRAWINGS SPOKE FOR THEMSELVES.

THE MOON WAS NOW ANOTHER WORLD.

IT HAD BEEN ONE ALL ALONG, OF COURSE, BUT NO ONE HAD REALIZED IT,

MERELY BECAUSE OUR EYES WEREN'T GOOD ENOUGH TO SEE.

AND NOT JUST THE MOON.

THROUGH HIS TELESCOPE GALILEO HAD DISCOVERED OTHER WONDERS: AN INFINITUDE OF STARS IN THE MILKY WAY

AND OTHER MOONS, CIRCLING JUPITER:

"FOUR PLANETS NEVER SEEN FROM THE BEGINNING OF THE WORLD RIGHT UP TO OUR DAY."

ALTHOUGH GALILEO'S BOOK DID NOT EXPLICITLY CHALLENGE CHURCH DOCTRINE, IT DID RAISE ALARMING QUESTIONS:

HOW MANY OTHER WORLDS *ARE* THERE?

AND IF THERE ARE OTHERS, DID GOD CREATE LIFE ON THEM AS WELL?

NOT EVERYONE BELIEVED THIS MESSAGE FROM THE STARS.

MANY HAVE SINCE PEERED THROUGH GALILEO'S TELESCOPE AND SEEN NO SIGN OF THESE NEW STARS.

PERHAPS HIS "DISCOVERY" IS ONLY AN ILLUSION CAUSED BY SPOTTY GLASS AND CONFUSED EYES?

IT SEEMED THAT GALILEO HAD MADE MORE ENEMIES THAN ALLIES.

WHAT HE NEEDED WAS THE VALIDATION OF SOMEONE WHOM NOBODY WOULD DOUBT.

WHAT HE NEEDED WAS THE IMPERIAL ASTRONOMER.

A GIFT, MASTER KEPLER, FROM THE DUKE OF TUSCANY...

KEPLER IMMEDIATELY WROTE A LETTER IN RESPONSE.

"TO THE NOBLE AND MOST EXCELLENT SIGNOR GALILEO GALILEI,

"I MAY PERHAPS SEEM RASH IN ACCEPTING YOUR CLAIMS SO READILY WITH NO SUPPORT FROM MY OWN EXPERIENCE.

"BY YOUR DISCOVERIES YOU CAUSED THE SUN OF TRUTH TO RISE, YOU ROUTED ALL THE GHOSTS OF PERPLEXITY TOGETHER WITH THEIR MOTHER, THE NIGHT, AND BY YOUR ACHIEVEMENT YOU SHOWED WHAT COULD BE DONE."

KEPLER DIDN'T YET HAVE HIS OWN TELESCOPE, SO HE COULDN'T SEE WHAT GALILEO SAW.

BUT HE COULD IMAGINE IT.

IN FACT, HE ALREADY HAD,

IN *THE DREAM*.

A YEAR BEFORE GALILEO'S DISCOVERY, KEPLER HAD DUSTED OFF HIS OLD DISSERTATION ABOUT THE MOON.

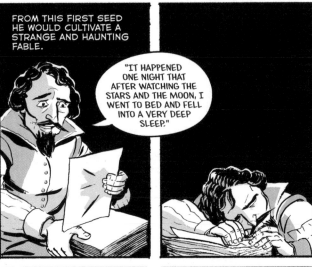

FROM THIS FIRST SEED HE WOULD CULTIVATE A STRANGE AND HAUNTING FABLE.

"IT HAPPENED ONE NIGHT THAT AFTER WATCHING THE STARS AND THE MOON, I WENT TO BED AND FELL INTO A VERY DEEP SLEEP."

KEPLER'S PURPOSE IN WRITING *THE DREAM* WAS TO PROVE WRONG ANYONE WHO DOUBTED THAT THE EARTH ORBITED THE SUN.

THIS, HOWEVER, WAS NO ORDINARY PROOF.

IT WAS AS MUCH A FAIRY TALE AS A TREATISE.

THE WORLD'S FIRST SCIENCE FICTION STORY.

KEPLER'S DREAM WOULD TAKE ALMOST FOUR CENTURIES TO COME TRUE.

ALTHOUGH HE WOULDN'T LIVE TO SEE THE DAY,

KEPLER WAS READY.

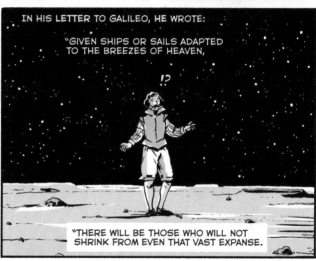

IN HIS LETTER TO GALILEO, HE WROTE:

"GIVEN SHIPS OR SAILS ADAPTED TO THE BREEZES OF HEAVEN,

"THERE WILL BE THOSE WHO WILL NOT SHRINK FROM EVEN THAT VAST EXPANSE.

"THEREFORE, FOR THE SAKE OF THOSE WHO, AS IT WERE, WILL PRESENTLY BE ON HAND TO ATTEMPT THIS VOYAGE,

"LET US ESTABLISH THE ASTRONOMY...

"...OF THE MOON."

45

CHAPTER V
COLUMBIA

SEVENTY MILES ABOVE HIS CREWMATES, MICHAEL COLLINS ORBITED THE MOON IN THE COMMAND MODULE.

HOUSTON, COLUMBIA.

HOW'S IT GOING?

MIKE, BE ADVISED WE HAVE AN UPDATE FOR YOU ON A POSSIBLE LOCATION OF THE LM.*

WHILE THE WORLD'S ATTENTION WAS TURNED TO THE MEN PREPARING TO STEP OUT ONTO THE LUNAR SURFACE, COLLINS KEPT A LOW PROFILE.

WE ESTIMATE THEY LANDED ABOUT FOUR MILES DOWNRANGE.

ROGER.

AS HE SAW IT, HIS JOB WAS TO "ACT LIKE A GOOD CHILD AND BE SEEN AND NOT HEARD."

*LUNAR MODULE, PRONOUNCED "LEM".

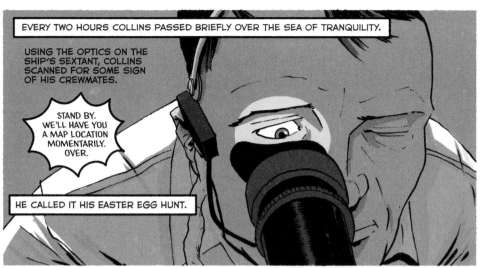

EVERY TWO HOURS COLLINS PASSED BRIEFLY OVER THE SEA OF TRANQUILITY.

USING THE OPTICS ON THE SHIP'S SEXTANT, COLLINS SCANNED FOR SOME SIGN OF HIS CREWMATES.

STAND BY. WE'LL HAVE YOU A MAP LOCATION MOMENTARILY. OVER.

HE CALLED IT HIS EASTER EGG HUNT.

COLLINS WAS ORBITING AT ABOUT 3700 MILES AN HOUR,

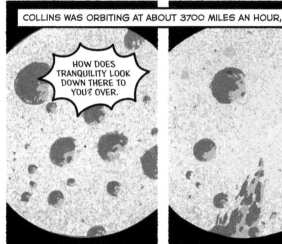

HOW DOES TRANQUILITY LOOK DOWN THERE TO YOU? OVER.

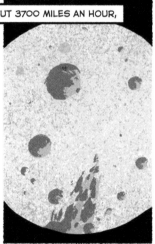

WELL, THE AREA LOOKS SMOOTH,

WHICH MEANT THAT WITH EACH PASS,

BUT I WAS UNABLE TO SEE THEM.

HE HAD JUST TWO MINUTES TO SEARCH FOR THE *EAGLE*.

MEANWHILE, IN THE MAPPING SCIENCES LABORATORY AT MISSION CONTROL, THE EXPERTS WERE STUMPED.

...FOR A TOTAL OF 14 POTENTIAL LANDING SITES.

AT LEAST THEY'RE ALL RELATIVELY CLOSE TOGETHER.

IN THE YEARS LEADING UP TO THE MOON LANDING, NASA HAD USED REMOTE PROBES TO PHOTOGRAPH AND SURVEY THE SEA OF TRANQUILITY.

GEOLOGISTS COMPILED THIS DATA INTO MAPS, WITH WHICH THE ASTRONAUTS COULD TRY TO MAKE SENSE OF THE MOON'S ALIEN LANDSCAPE.

DO YOU HAVE ANY TOPOGRAPHICAL CLUES THAT MIGHT HELP ME OUT HERE?

BUT ULTIMATELY THE ONLY WAY FOR COLLINS TO FIGURE OUT WHERE HIS CREWMATES HAD LANDED WAS OLD-FASHIONED DEAD RECKONING.

THE BEST WE CAN DO IS ADVISE YOU TO LOOK TO THE WEST OF THE IRREGULARLY SHAPED CRATER, THEN WORK ON DOWN TO THE SOUTHWEST OF IT...

NEVER BEFORE IN HISTORY HAD HUMANS MARSHALLED SUCH A CONCENTRATION OF TECHNICAL PROWESS.

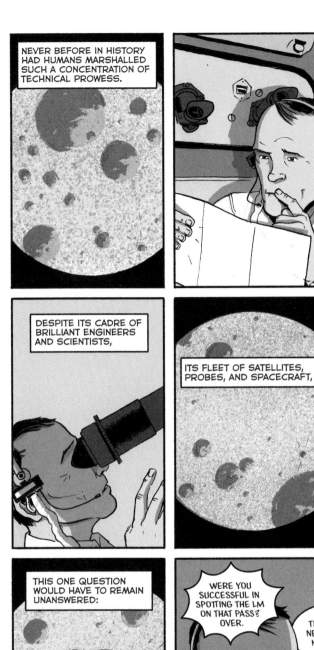

AND YET, DESPITE THE VAST RESOURCES AT NASA'S DISPOSAL,

DESPITE ITS CADRE OF BRILLIANT ENGINEERS AND SCIENTISTS,

ITS FLEET OF SATELLITES, PROBES, AND SPACECRAFT,

AND ITS GLOBE-SPANNING COMMUNICATIONS NETWORK,

THIS ONE QUESTION WOULD HAVE TO REMAIN UNANSWERED:

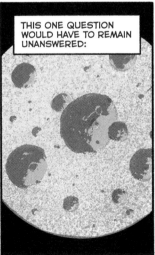

WERE YOU SUCCESSFUL IN SPOTTING THE LM ON THAT PASS? OVER.

THAT'S A NEGATIVE. NO JOY.

WHERE WERE THEY?

CHAPTER VI
STORYTELLERS

THERE ARE MANY THINGS TO PREVENT US FROM LEAVING
OUR OWN WORLD AND GOING TO THE MOON. TO CONSOLE
OURSELVES LET US GUESS ALL WE CAN ABOUT IT.

--BERNARD LE BOVIER DE FONTENELLE

PALACE OF WHITEHALL, LONDON, 1662

THE COURT OF CHARLES II

HIS MAJESTY WILL SEE YOU NOW.

MR. WREN, WHAT WONDERS HAVE YOU BROUGHT TO ME TODAY?

YOUR MAJESTY, I DELIVER TO YOU, AS PROMISED, *THE MOON*.

OR AS CLOSE A FACSIMILE AS HAS YET BEEN RENDERED.

A TRIUMPH OF ARTIFICE! EVERY HILL, EVERY CAVITY...

BUT--HOW DO I POSE THIS WITHOUT SEEMING IGNORANT--

I FEAR, MR. WREN, THAT YOUR GLOBE IS UNFINISHED.

WHY, ONE WHOLE SIDE OF IT IS...*EMPTY.*

AH, I SEE. WELL, TO PUT IT BLUNTLY, YOUR MAJESTY, THE MOON ITSELF IS TO BLAME.

THE FAR SIDE OF HER FACE IS *ALWAYS* HIDDEN FROM OUR VIEW.

51

WHEN THE MOON WAS FIRST CREATED, IT MOVED MUCH FASTER THAN IT DOES NOW.

OVER EONS, THE GRAVITATIONAL TUG OF THE EARTH WORKED LIKE A COSMIC BRAKE, SLOWING THE MOON IN ITS ROTATION,

SO THAT NOW IT TAKES ABOUT 27 DAYS FOR THE MOON TO SPIN ONCE.

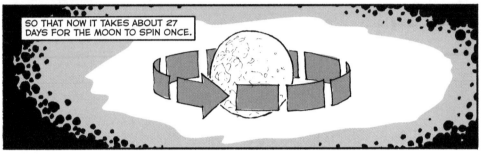

IN OTHER WORDS, ONE LUNAR DAY...

...IS ABOUT EQUAL TO 27 EARTH DAYS.

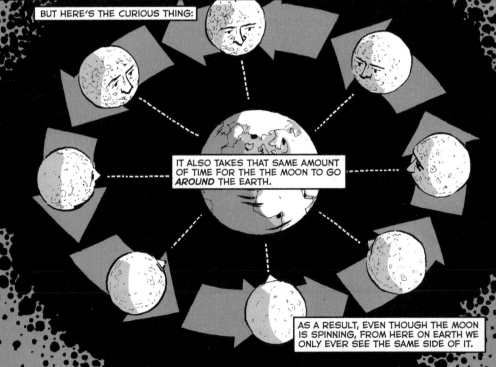

BUT HERE'S THE CURIOUS THING:

IT ALSO TAKES THAT SAME AMOUNT OF TIME FOR THE THE MOON TO GO *AROUND* THE EARTH.

AS A RESULT, EVEN THOUGH THE MOON IS SPINNING, FROM HERE ON EARTH WE ONLY EVER SEE THE SAME SIDE OF IT.

THE INVISIBLE FAR SIDE OF THE MOON WAS LIKE THE BLANK SPACES ON THE MARGINS OF EARLY MAPS.

TERRA INCOGNITA

A FERTILE PLACE FOR ALL SORTS OF FANTASTIC IDEAS.

WRITERS FROM ANTIQUITY TO THE RENAISSANCE IMAGINED WHAT THEY MIGHT FIND UP THERE, FAR BEYOND THE BORDERS OF OUR OWN KNOWN WORLD.

THREE-HEADED HORSE VULTURES.

A RACE OF GIANTS WITH LETTUCE-LEAF TAILS AND REMOVABLE EYEBALLS.

SNAKELIKE CREATURES THAT HATCH FROM COCOONS AT NIGHT.

A PLACE WHERE MURDER IS IMPOSSIBLE BECAUSE ALL WOUNDS ARE CURABLE.

WHERE THE OLD ARE MADE YOUNG.

WHERE ALL THE LOST THINGS ON EARTH WASH UP.

53

AS FANTASTIC AS THESE LUNAR TALES MAY SEEM, THEY WEREN'T ANY STRANGER THAN THE STORIES TOLD BY SAILORS RETURNING FROM THE FAR CORNERS OF THE GLOBE.

DURING THE 16TH AND 17TH CENTURIES, EUROPEAN EXPLORERS VENTURED TO LANDS THAT PREVIOUS GENERATIONS HAD SIMPLY CALLED *TERRA INCOGNITA*--THE UNKNOWN.

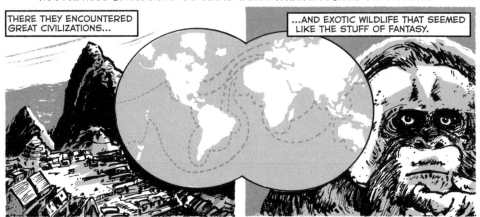

THERE THEY ENCOUNTERED GREAT CIVILIZATIONS...

...AND EXOTIC WILDLIFE THAT SEEMED LIKE THE STUFF OF FANTASY.

FOR ANYONE RAISED ON THESE STORIES OF DISCOVERY, IT DIDN'T SEEM OUTLANDISH TO THINK THAT IF ONLY WE HAD THE RIGHT KIND OF SHIP...

FRANCIS GODWIN, 1638.

CYRANO DE BERGERAC, 1657.

DAVID RUSSEN, 1703.

...WE MIGHT SOMEDAY SAIL TO THE *TERRA INCOGNITA* OF THE MOON.

IN 1687 ONE BOOK FOREVER CHANGED THE CONVERSATION ABOUT SPACEFLIGHT. FOR THE FIRST TIME, A VOYAGE TO THE MOON BECAME AT LEAST PLAUSIBLE, IF NOT YET POSSIBLE.

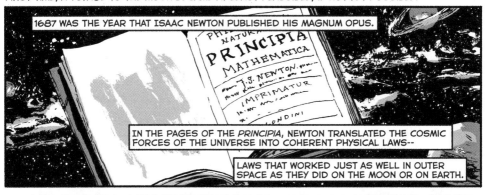

1687 WAS THE YEAR THAT ISAAC NEWTON PUBLISHED HIS MAGNUM OPUS.

IN THE PAGES OF THE *PRINCIPIA*, NEWTON TRANSLATED THE COSMIC FORCES OF THE UNIVERSE INTO COHERENT PHYSICAL LAWS--

LAWS THAT WORKED JUST AS WELL IN OUTER SPACE AS THEY DID ON THE MOON OR ON EARTH.

IN ONE OF THESE LAWS NEWTON DEVELOPED AN IDEA FIRST PROPOSED BY GALILEO AND KEPLER: *GRAVITY*.

GALILEO HAD OBSERVED THE WAY THAT OBJECTS ON EARTH FALL. FROM THIS HE DEVISED A WAY TO PREDICT THE PATHS OF CANNONBALLS.

AND REMEMBER, KEPLER WAS THE ONE WHO FIRST FIGURED OUT THE ORBITS OF THE PLANETS AROUND THE SUN.

NEWTON'S GREAT INSIGHT WAS TO SHOW HOW THE FORCE THAT EXPLAINED THE MOTION OF GALILEO'S CANNONBALLS AND KEPLER'S PLANETS WAS THE SAME.

BY CONSIDERING THE MOTIONS OF PROJECTILES, I SHALL EASILY DEMONSTRATE HOW THE PLANETS MAINTAIN THEIR ORBITS.

TO ILLUSTRATE HIS POINT, NEWTON PROPOSED A THOUGHT EXPERIMENT:

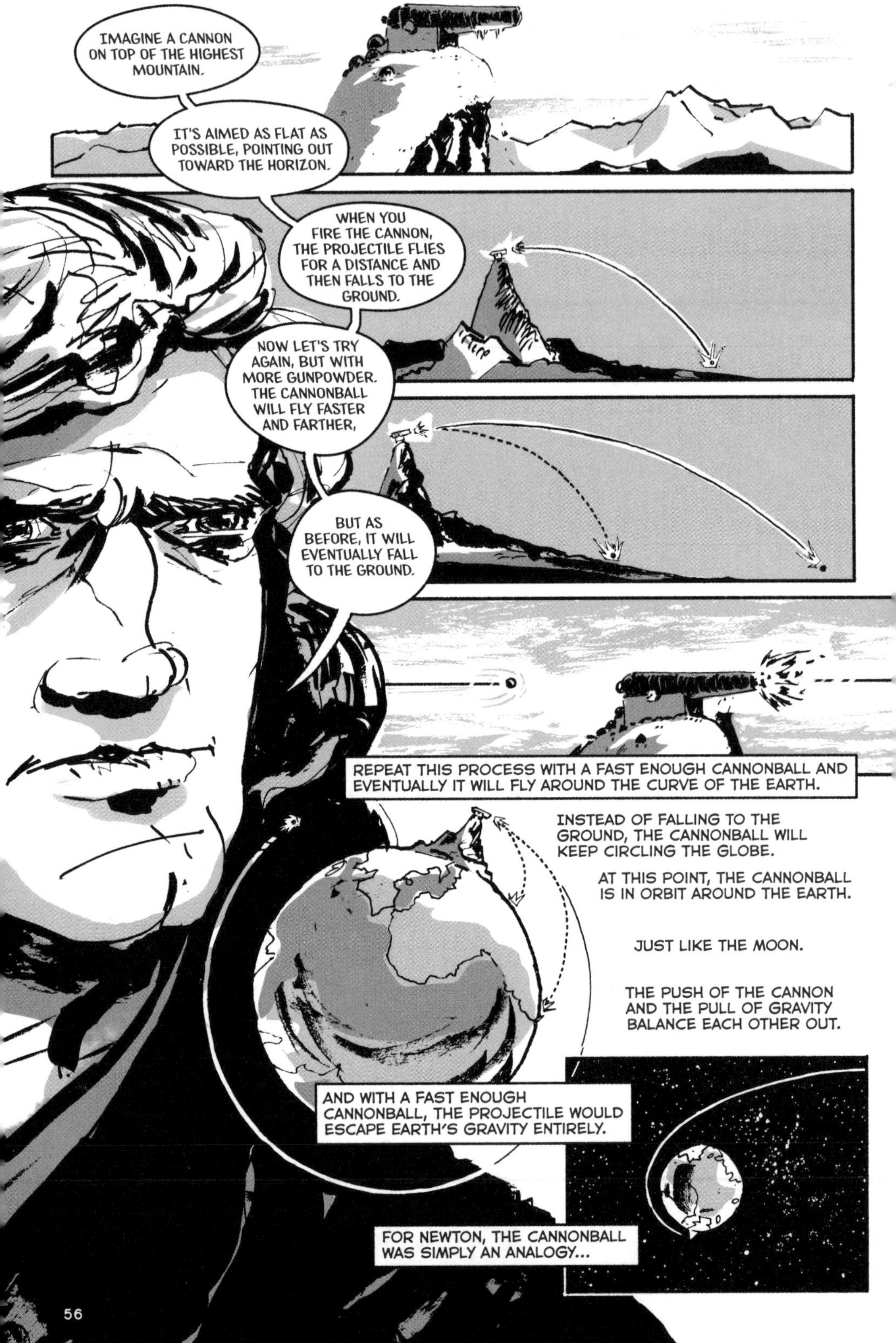
IMAGINE A CANNON ON TOP OF THE HIGHEST MOUNTAIN.
IT'S AIMED AS FLAT AS POSSIBLE, POINTING OUT TOWARD THE HORIZON.
WHEN YOU FIRE THE CANNON, THE PROJECTILE FLIES FOR A DISTANCE AND THEN FALLS TO THE GROUND.
NOW LET'S TRY AGAIN, BUT WITH MORE GUNPOWDER. THE CANNONBALL WILL FLY FASTER AND FARTHER,
BUT AS BEFORE, IT WILL EVENTUALLY FALL TO THE GROUND.
REPEAT THIS PROCESS WITH A FAST ENOUGH CANNONBALL AND EVENTUALLY IT WILL FLY AROUND THE CURVE OF THE EARTH.
INSTEAD OF FALLING TO THE GROUND, THE CANNONBALL WILL KEEP CIRCLING THE GLOBE.
AT THIS POINT, THE CANNONBALL IS IN ORBIT AROUND THE EARTH.
JUST LIKE THE MOON.
THE PUSH OF THE CANNON AND THE PULL OF GRAVITY BALANCE EACH OTHER OUT.
AND WITH A FAST ENOUGH CANNONBALL, THE PROJECTILE WOULD ESCAPE EARTH'S GRAVITY ENTIRELY.
FOR NEWTON, THE CANNONBALL WAS SIMPLY AN ANALOGY...

...BUT FOR JULES VERNE, WRITING IN PARIS IN 1865, IT WAS A REVELATION.

"...AND TO TELL THE TRUTH, THE PLANETS THEMSELVES ARE JUST PROJECTILES,

"SIMPLY CANNONBALLS FIRED BY THE HAND OF OUR CREATOR."

FOR THE PAST FOUR YEARS, VERNE HAD BEEN FOLLOWING NEWS OF THE CIVIL WAR IN AMERICA.

BOTH SIDES HAD INVENTED POWERFUL GUNS AND ARTILLERY, INFLICTING UNIMAGINABLE CARNAGE.

BUT THE QUESTION THAT FASCINATED VERNE NOW THAT THE WAR WAS OVER:

WHAT WOULD BECOME OF THESE AMERICAN GENIUSES,

THE ONES WHO BUILT THE GUNS THAT TORE THE COUNTRY APART?

SO HE DECIDED TO WRITE A BOOK ABOUT IT.

TO SET THE SCENE: OCTOBER 5, 1865.

GALLANT COLLEAGUES...

A SPECIAL MEETING OF THE BALTIMORE GUN CLUB

...MANY YEARS WILL GO BY BEFORE OUR CANNONS THUNDER AGAIN ON THE BATTLEFIELD.

FOR THE PAST FEW MONTHS, MY GALLANT COLLEAGUES,

I'VE WONDERED IF WE COULD STAY IN OUR FIELD OF EXPERTISE YET UNDERTAKE SOME GREAT EXPERIMENT WORTHY OF THE 19TH CENTURY.

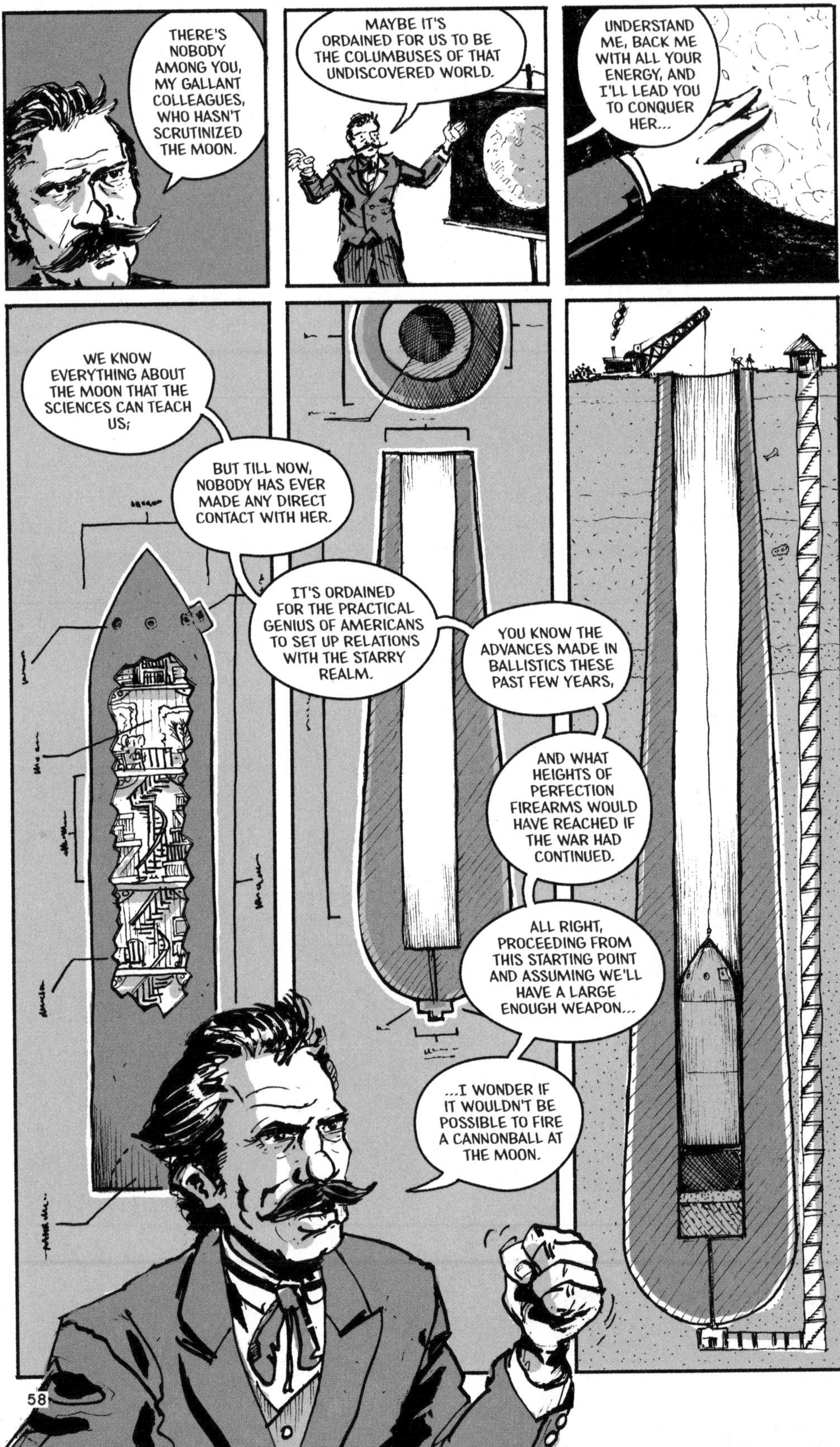
THERE'S NOBODY AMONG YOU, MY GALLANT COLLEAGUES, WHO HASN'T SCRUTINIZED THE MOON.
MAYBE IT'S ORDAINED FOR US TO BE THE COLUMBUSES OF THAT UNDISCOVERED WORLD.
UNDERSTAND ME, BACK ME WITH ALL YOUR ENERGY, AND I'LL LEAD YOU TO CONQUER HER...
WE KNOW EVERYTHING ABOUT THE MOON THAT THE SCIENCES CAN TEACH US;
BUT TILL NOW, NOBODY HAS EVER MADE ANY DIRECT CONTACT WITH HER.
IT'S ORDAINED FOR THE PRACTICAL GENIUS OF AMERICANS TO SET UP RELATIONS WITH THE STARRY REALM.
YOU KNOW THE ADVANCES MADE IN BALLISTICS THESE PAST FEW YEARS,
AND WHAT HEIGHTS OF PERFECTION FIREARMS WOULD HAVE REACHED IF THE WAR HAD CONTINUED.
ALL RIGHT, PROCEEDING FROM THIS STARTING POINT AND ASSUMING WE'LL HAVE A LARGE ENOUGH WEAPON...
...I WONDER IF IT WOULDN'T BE POSSIBLE TO FIRE A CANNONBALL AT THE MOON.

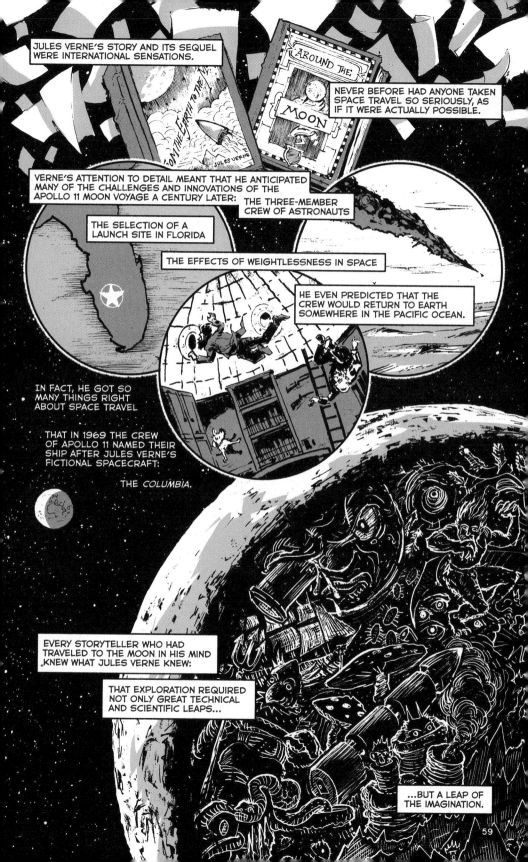

JULES VERNE'S STORY AND ITS SEQUEL WERE INTERNATIONAL SENSATIONS.

AROUND THE MOON

ON THE EARTH TO THE M... JULES VERNE

NEVER BEFORE HAD ANYONE TAKEN SPACE TRAVEL SO SERIOUSLY, AS IF IT WERE ACTUALLY POSSIBLE.

VERNE'S ATTENTION TO DETAIL MEANT THAT HE ANTICIPATED MANY OF THE CHALLENGES AND INNOVATIONS OF THE APOLLO 11 MOON VOYAGE A CENTURY LATER:

THE THREE-MEMBER CREW OF ASTRONAUTS

THE SELECTION OF A LAUNCH SITE IN FLORIDA

THE EFFECTS OF WEIGHTLESSNESS IN SPACE

HE EVEN PREDICTED THAT THE CREW WOULD RETURN TO EARTH SOMEWHERE IN THE PACIFIC OCEAN.

IN FACT, HE GOT SO MANY THINGS RIGHT ABOUT SPACE TRAVEL

THAT IN 1969 THE CREW OF APOLLO 11 NAMED THEIR SHIP AFTER JULES VERNE'S FICTIONAL SPACECRAFT:

THE COLUMBIA.

EVERY STORYTELLER WHO HAD TRAVELED TO THE MOON IN HIS MIND KNEW WHAT JULES VERNE KNEW:

THAT EXPLORATION REQUIRED NOT ONLY GREAT TECHNICAL AND SCIENTIFIC LEAPS...

...BUT A LEAP OF THE IMAGINATION.

CHAPTER VII

EAGLE

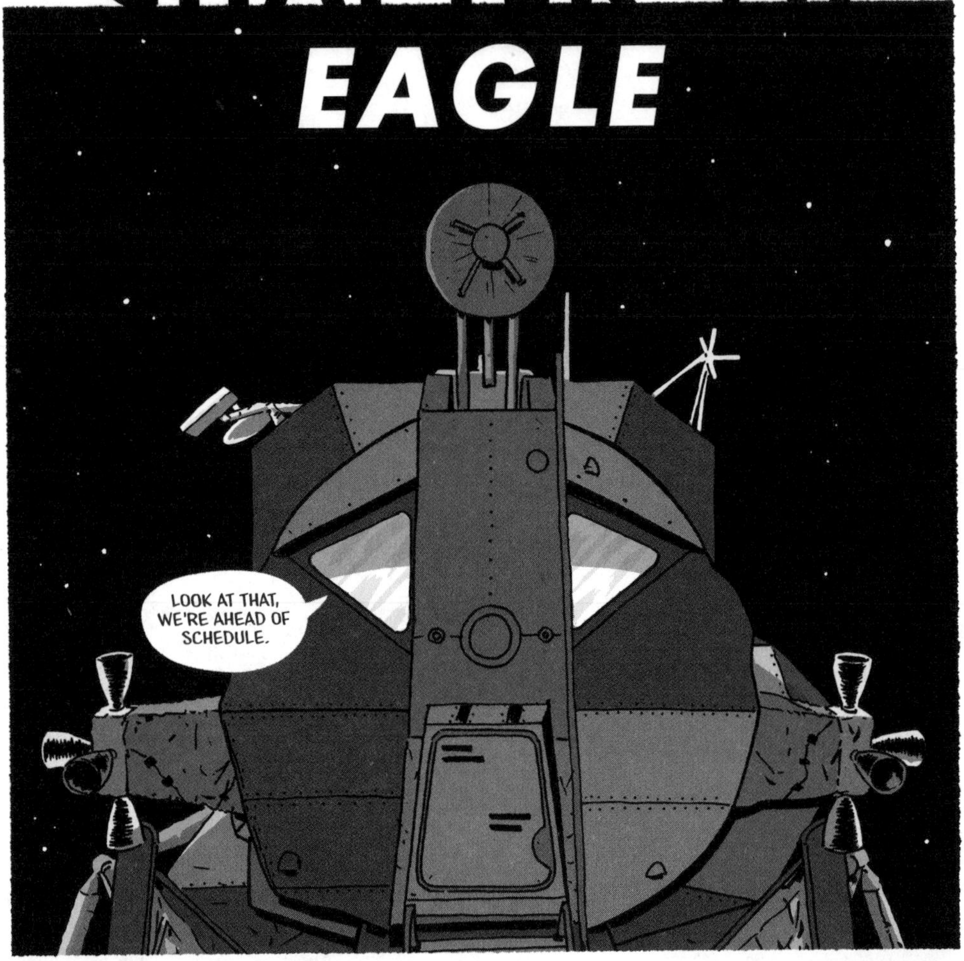
LOOK AT THAT, WE'RE AHEAD OF SCHEDULE.

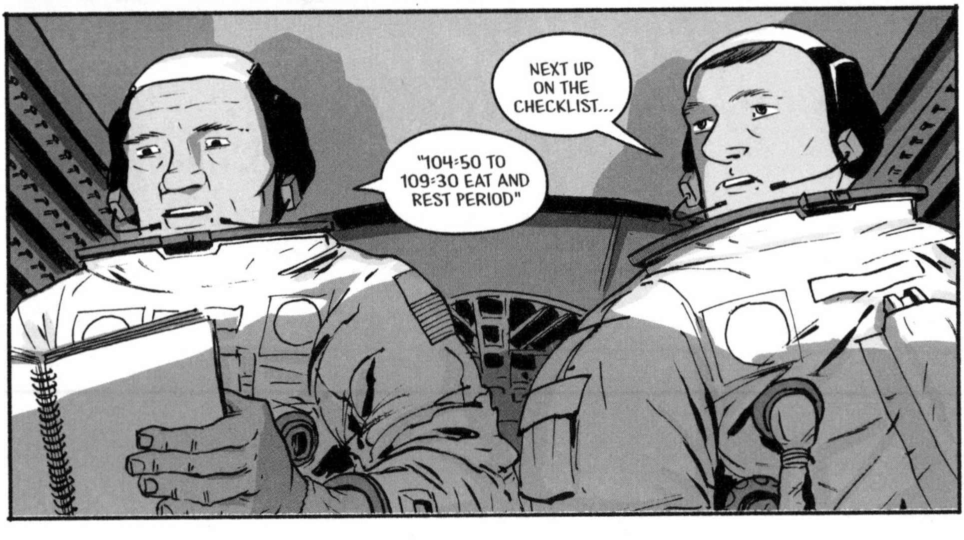
NEXT UP ON THE CHECKLIST...
"104:50 TO 109:30 EAT AND REST PERIOD"

*EXTRAVEHICULAR ACTIVITY, AKA MOONWALK

YOU GUYS WILL BE ON PRIME-TIME TV THERE.

HOPE OUR CAMERA WORKS.

YOU WANT THE BEEF STEW OR THE CHICKEN SOUP?

SOUP'S FINE. YOU CAN JUST SET IT THERE.

THERE'S SOMETHING I NEED TO DO FIRST.

HOUSTON, TRANQUILITY. OVER.

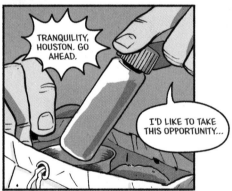

TRANQUILITY, HOUSTON. GO AHEAD.

I'D LIKE TO TAKE THIS OPPORTUNITY...

...TO ASK EVERY PERSON LISTENING IN...

...WHOEVER AND WHEREVER THEY MAY BE....

...TO PAUSE FOR A MOMENT AND CONTEMPLATE THE EVENTS OF THE PAST FEW HOURS...

...AND TO GIVE THANKS IN HIS OR HER OWN WAY. OVER.

THEN ALDRIN MUTED HIS MICROPHONE.

SOFTLY, HE SPOKE THE WORDS THAT HIS PASTOR HAD PREPARED FOR HIM.

"I AM THE VINE, YOU ARE THE BRANCHES.

WHOEVER REMAINS IN ME, AND I IN HIM,

WILL BEAR MUCH FRUIT; FOR YOU CAN DO NOTHING WITHOUT ME."

ALDRIN

63

CHAPTER VIII
ROCKETS

THE EARTH IS THE CRADLE OF REASON, BUT ONE CANNOT
LIVE IN THE CRADLE FOREVER.

--KONSTANTIN TSIOLKOVSKY

ONCE THE ROCKETS ARE UP,
WHO CARES WHERE THEY COME DOWN
THAT'S NOT MY DEPARTMENT,
SAYS WERNHER VON BRAUN.

--TOM LEHRER

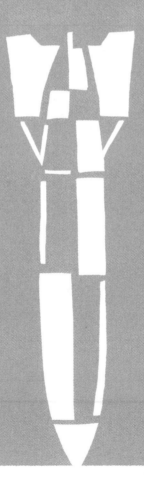

APRIL 10, 1945

GERMANY IS ON THE BRINK OF COLLAPSE.

BY THE END OF THE MONTH, HITLER WILL BE DEAD.

AS THE ALLIES ADVANCE TOWARD BERLIN FROM EAST AND WEST...

...THEY UNCOVER THE AFTERMATH OF INDESCRIBABLE ATROCITIES.

POMOCY! POMÓŻCIE NAM.

SZYBKO, WYŁAMCIE ZAMEK!

BASH

WHAT'S HE SAYING?

I GOT NO CLUE...

CHODŹCIE, CHODŹCIE, ZA MNĄ!

WHAT PRIVATE JOHN GALIONE FOUND THAT DAY WOULD HAUNT HIM FOR THE REST OF HIS LIFE.

"WE THOUGHT NOTHING COULD HURT US. WE WERE HARD FROM WAR. BUT WHEN WE WALKED IN THERE, WE COULDN'T...

WHAT IS THIS PLACE?

CHODŹ...

"...WE COULDN'T BELIEVE ANY HUMAN BEING COULD BE SO CRUEL."

TUTAJ. MUSISZ TO ZOBACZYĆ.

FLIK

WHAT? I CAN'T UNDERSTAND...

RAKIETY!

RAKIETY!

ROCKETS.

MEANWHILE, A FEW HUNDRED MILES AWAY, IN THE GERMAN ALPS.

I'LL TAKE ANOTHER SLICE OF TOAST.

THE U.S. ARMY HAS FOUND A GROUP OF NAZI SCIENTISTS HIDING OUT IN A VILLAGE.

IF I'D BEEN GIVEN TWO MORE YEARS...PASS THE JAM...

...TWO MORE YEARS AND MY V-2 ROCKET WOULD HAVE WON THE WAR FOR GERMANY.

AT FIRST, NONE OF THE SOLDIERS REALIZED WHOM THEY HAD FOUND.

MY NAME IS WERNHER VON BRAUN, AND I WOULD LIKE TO SPEAK TO GENERAL EISENHOWER.

IF IT SEEMED LIKE AN AUDACIOUS REQUEST,

THAT'S ONLY BECAUSE FEW PEOPLE YET KNEW JUST HOW AUDACIOUS THIS MAN COULD BE.

VON BRAUN HAD BEEN AN OFFICER IN THE NAZI SS

AND THE LEAD ENGINEER OF HITLER'S SECRET ROCKET PROGRAM.

AND SOON HE WOULD BECOME AN AMERICAN CITIZEN.

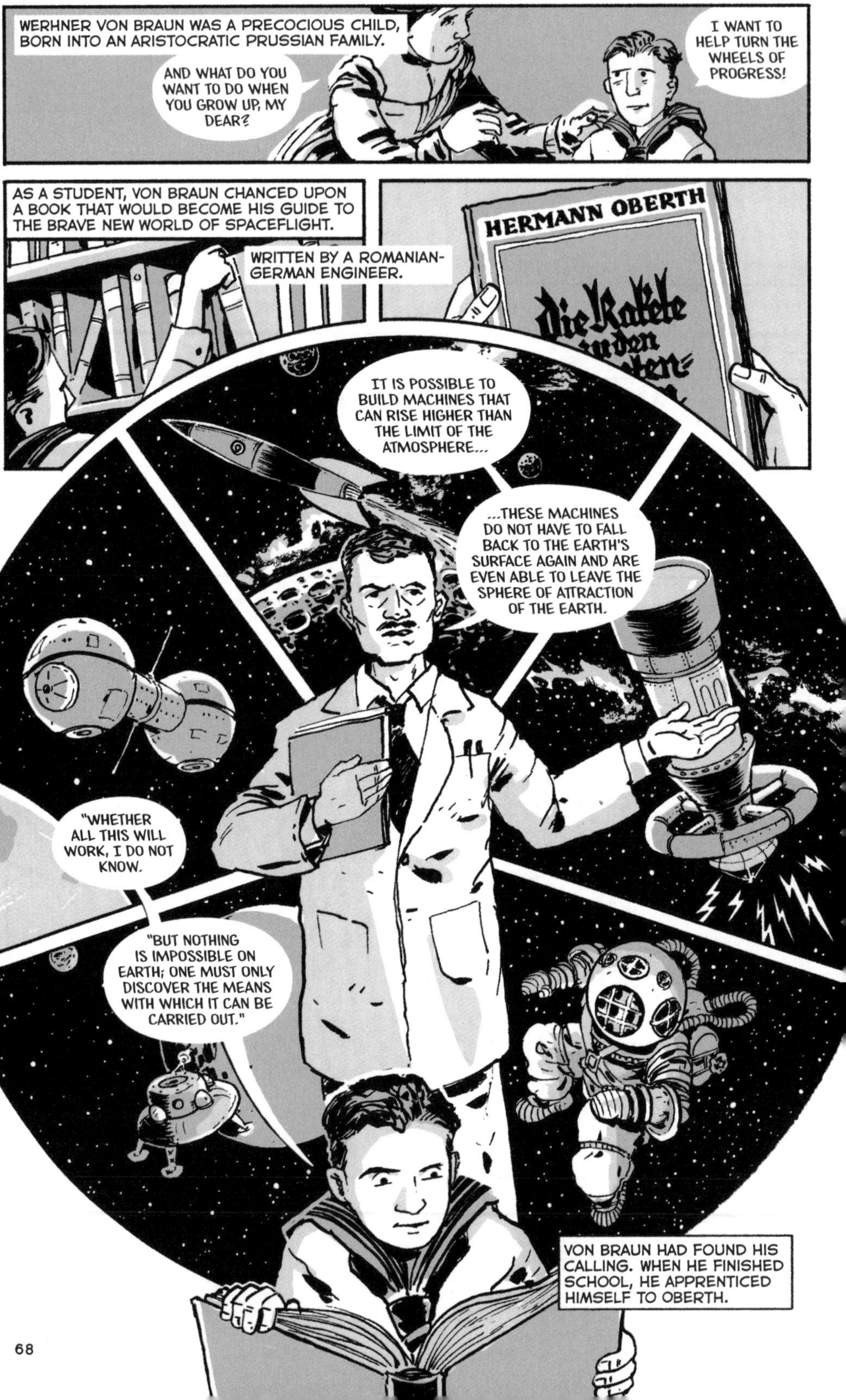
WERHNER VON BRAUN WAS A PRECOCIOUS CHILD, BORN INTO AN ARISTOCRATIC PRUSSIAN FAMILY.
AND WHAT DO YOU WANT TO DO WHEN YOU GROW UP, MY DEAR?
I WANT TO HELP TURN THE WHEELS OF PROGRESS!
AS A STUDENT, VON BRAUN CHANCED UPON A BOOK THAT WOULD BECOME HIS GUIDE TO THE BRAVE NEW WORLD OF SPACEFLIGHT.
WRITTEN BY A ROMANIAN-GERMAN ENGINEER.
HERMANN OBERTH
IT IS POSSIBLE TO BUILD MACHINES THAT CAN RISE HIGHER THAN THE LIMIT OF THE ATMOSPHERE...
...THESE MACHINES DO NOT HAVE TO FALL BACK TO THE EARTH'S SURFACE AGAIN AND ARE EVEN ABLE TO LEAVE THE SPHERE OF ATTRACTION OF THE EARTH.
"WHETHER ALL THIS WILL WORK, I DO NOT KNOW.
"BUT NOTHING IS IMPOSSIBLE ON EARTH; ONE MUST ONLY DISCOVER THE MEANS WITH WHICH IT CAN BE CARRIED OUT."
VON BRAUN HAD FOUND HIS CALLING. WHEN HE FINISHED SCHOOL, HE APPRENTICED HIMSELF TO OBERTH.

IN HINDSIGHT, HERMANN OBERTH MAY SEEM LIKE A VISIONARY,

BUT AT THE TIME, HIS PLANS FOR A ROCKET BIG ENOUGH TO CARRY A MAN INTO SPACE SEEMED LAUGHABLE.

ROCKETS WERE SMALL AND DECIDEDLY EXPLOSIVE.

WHICH WAS EXACTLY THE POINT OF THE FIRST ROCKETS, INVENTED IN CHINA AROUND THE 13TH CENTURY.

THESE "FIRE ARROWS" DID DOUBLE DUTY AS WEAPONS OF WAR AND FIREWORKS FOR RELIGIOUS HOLIDAYS,

SUCH AS THE MID-AUTUMN FESTIVAL IN HONOR OF THE MOON GODDESS CHANG'E.

FROM ITS EARLIEST INCARNATION, THE ROCKET HAS ALWAYS BEEN AN AMBIVALENT INVENTION:

A TOOL OF DEVASTATION.

A SYMBOL OF TERROR.

A SOURCE OF WONDER.

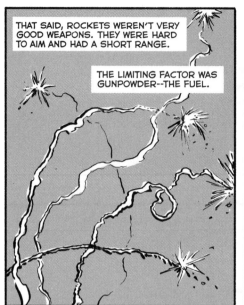

THAT SAID, ROCKETS WEREN'T VERY GOOD WEAPONS. THEY WERE HARD TO AIM AND HAD A SHORT RANGE.

THE LIMITING FACTOR WAS GUNPOWDER--THE FUEL.

ACTUALLY, WHEN WE'RE TALKING ABOUT ROCKETS, "FUEL" ISN'T QUITE RIGHT.

A CAR ENGINE RUNS BY BURNING GASOLINE.

THAT'S THE FUEL.

BUT GASOLINE CAN'T BURN WITHOUT OXYGEN, WHICH IS WHY CAR ENGINES NEED AIR.

NEED $ 4 AIR

GUNPOWDER, HOWEVER, IS A *MIXTURE* OF CHEMICALS. IT'S THE FUEL AND THE OXYGEN COMBINED. WHICH MEANS IT DOESN'T NEED ANY OTHER CHEMICALS FOR IT TO BURN.

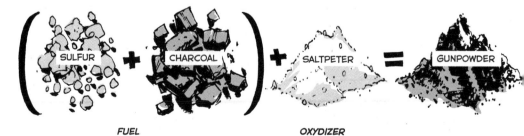

(SULFUR + CHARCOAL) + SALTPETER = GUNPOWDER

FUEL   *OXYDIZER*

IN THE WORLD OF ROCKETRY, SUCH A MIXTURE IS CALLED A *PROPELLANT*.

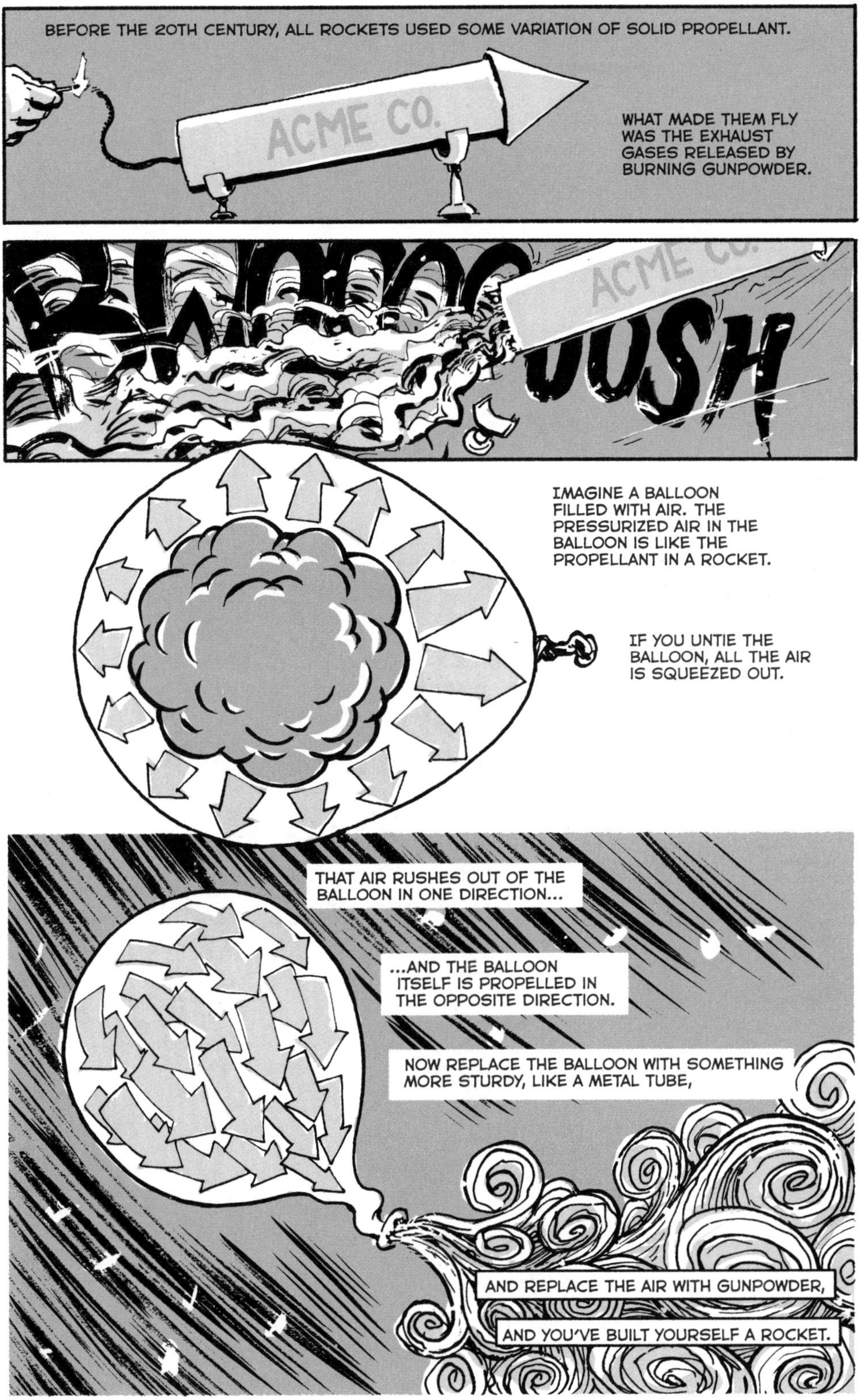

BEFORE THE 20TH CENTURY, ALL ROCKETS USED SOME VARIATION OF SOLID PROPELLANT.
ACME CO.
WHAT MADE THEM FLY WAS THE EXHAUST GASES RELEASED BY BURNING GUNPOWDER.
ACME CO.
IMAGINE A BALLOON FILLED WITH AIR. THE PRESSURIZED AIR IN THE BALLOON IS LIKE THE PROPELLANT IN A ROCKET.
IF YOU UNTIE THE BALLOON, ALL THE AIR IS SQUEEZED OUT.
THAT AIR RUSHES OUT OF THE BALLOON IN ONE DIRECTION...
...AND THE BALLOON ITSELF IS PROPELLED IN THE OPPOSITE DIRECTION.
NOW REPLACE THE BALLOON WITH SOMETHING MORE STURDY, LIKE A METAL TUBE,
AND REPLACE THE AIR WITH GUNPOWDER,
AND YOU'VE BUILT YOURSELF A ROCKET.

BUT GUNPOWDER WON'T GET YOU TO THE MOON.

THE FIRST PERSON TO FIGURE OUT WHY NOT WAS A RUSSIAN NAMED KONSTANTIN TSIOLKOVSKY, BORN IN 1857.

WHEN HE WASN'T BUSY TEACHING HIGH SCHOOL MATH OR READING JULES VERNE, HE WAS CALCULATING HOW TO SEND A ROCKET INTO SPACE.

THERE ARE TWO THINGS WE NEED TO DECIDE BEFORE WE CAN SEND SOMEONE TO SPACE IN A ROCKET:

WHERE DO WE WANT HIM TO GO?

AND WHAT PROPELLANT ARE WE GOING TO USE?

IF WE ANSWER THESE TWO QUESTIONS, THESE TWO VARIABLES, MY EQUATION WILL TELL US HOW TO DESIGN THE ROCKET.

SAY, FOR EXAMPLE, THAT WE WANT TO GO FROM THE EARTH TO THE MOON

AND THAT WE'RE GOING TO USE GUNPOWDER TO PROPEL OUR ROCKET.

HMMM...NOT GOOD.

TSIOLKOVSKY'S EQUATION TELLS US THAT WE'RE GOING TO HAVE A PROBLEM.

ACCORDING TO MY CALCULATIONS, THE GUNPOWDER IS GOING TO TAKE UP 96 PERCENT OF THE ROCKET'S MASS.

IN OTHER WORDS, NEARLY ALL OF OUR ROCKET IS GOING TO HAVE TO BE PROPELLANT.

EVERYTHING ELSE--

THE ASTRONAUTS,

ALL THE STUFF THAT KEEPS THE ASTRONAUTS ALIVE,

THE METAL THAT THE ROCKET IS MADE OF--

HAS TO FIT IN THE REMAINING FOUR PERCENT OF THE TOTAL MASS.

CLEARLY, IF WE WISH TO SEND A MAN TO SPACE, GUNPOWDER IS OUT OF THE QUESTION.

TSIOLKOVSKY SOLVED THIS PROBLEM BY PROPOSING A LIQUID PROPELLANT,

A VOLATILE COMBINATION OF OXYGEN AND HYDROGEN THAT IS SIGNIFICANTLY MORE POWERFUL THAN GUNPOWDER.

AND TO EXTEND THE RANGE OF THE ROCKET, WHAT IF WE STACK SEVERAL ENGINES ON TOP OF EACH OTHER, IGNITING THEM IN STAGES...

HE PUBLISHED HIS WORK IN 1903, HOPING TO CONVINCE THE RUSSIAN GOVERNMENT THAT SPACEFLIGHT COULD BE A REALITY.

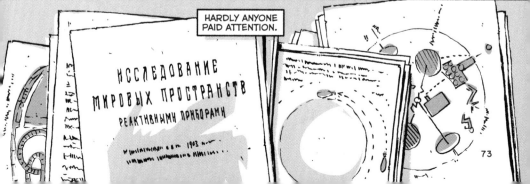

HARDLY ANYONE PAID ATTENTION.

ИССЛЕДОВАНИЕ МИРОВЫХ ПРОСТРАНСТВ РЕАКТИВНЫМИ ПРИБОРАМИ

73

ROBERT GODDARD HAD NEVER HEARD OF TSIOLKOVSKY--

IN FACT, HE HADN'T EVEN GRADUATED FROM HIGH SCHOOL--

YET HE TOO SHARED THE DREAM.

IN 1899, WHILE TSIOLKOVSKY WAS IMAGINING EVER-GRANDER ROCKET SHIPS,

GODDARD CLIMBED TO THE UPPER LIMBS OF A CHERRY TREE AT HIS HOUSE IN WORCESTER, MASSACHUSETTS, AND DREAMED OF OUTER SPACE.

"I WAS A DIFFERENT BOY WHEN I DESCENDED THE TREE FROM WHEN I ASCENDED, FOR EXISTENCE AT LAST SEEMED VERY PURPOSIVE."

FIFTEEN YEARS LATER, ARMED WITH AN ADVANCED DEGREE IN PHYSICS, GODDARD HAD BECOME OBSESSED WITH ROCKETS.

"YOUR APPLICATION FOR U.S. PATENT NO. 1,102,653, 'ROCKET APPARATUS,' HAS BEEN APPROVED..."

GODDARD HAD INDEPENDENTLY REACHED THE SAME CONCLUSIONS AS TSIOLKOVSKY:

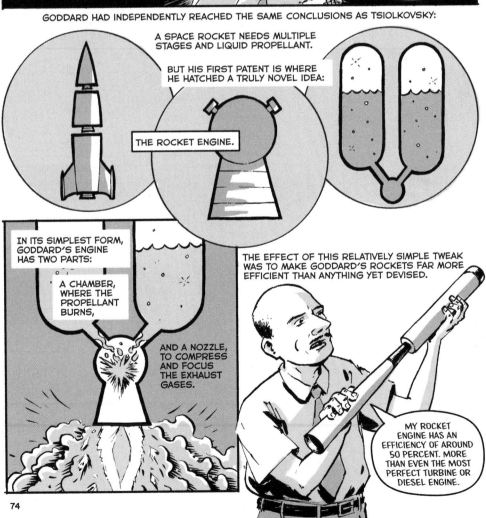

A SPACE ROCKET NEEDS MULTIPLE STAGES AND LIQUID PROPELLANT.

BUT HIS FIRST PATENT IS WHERE HE HATCHED A TRULY NOVEL IDEA:

THE ROCKET ENGINE.

IN ITS SIMPLEST FORM, GODDARD'S ENGINE HAS TWO PARTS:

A CHAMBER, WHERE THE PROPELLANT BURNS,

AND A NOZZLE, TO COMPRESS AND FOCUS THE EXHAUST GASES.

THE EFFECT OF THIS RELATIVELY SIMPLE TWEAK WAS TO MAKE GODDARD'S ROCKETS FAR MORE EFFICIENT THAN ANYTHING YET DEVISED.

MY ROCKET ENGINE HAS AN EFFICIENCY OF AROUND 50 PERCENT. MORE THAN EVEN THE MOST PERFECT TURBINE OR DIESEL ENGINE.

IN 1919, GODDARD WROTE UP THE RESULTS OF HIS EXPERIMENTS FOR AN OBSCURE TECHNICAL JOURNAL PUBLISHED BY THE SMITHSONIAN INSTITUTION.

IT PROBABLY WOULD HAVE BEEN IGNORED IF IT WEREN'T FOR A BRIEF MENTION OF THE MOON AT THE END OF THE ESSAY.

Herald

SCIENTIST TO SHOOT MOON WITH ROCKE

Daily Sun

SAVANT INVENTS MOON ROCKET

Tribune

A MODERN JULES VERNE

NEWSPAPERS AROUND THE COUNTRY THRILLED AT THIS UNKNOWN PROFESSOR WHO WANTED TO "SHOOT THE MOON."

THE NEW YORK TIMES, IN PARTICULAR, WAS SKEPTICAL. IT WAS "ABSURD" TO THINK THAT A ROCKET COULD WORK IN THE VACUUM OF SPACE.

"TO CLAIM THAT IT WOULD... IS TO DENY A FUNDAMENTAL LAW OF DYNAMICS, AND ONLY DR. EINSTEIN AND HIS CHOSEN DOZEN...ARE LICENSED TO DO THAT."

REGARDLESS, EAGER READERS VOLUNTEERED TO BE PASSENGERS ON A JOURNEY THAT THEY ASSUMED WAS IMMINENT.

I'M READY!

LITTLE DID THEY KNOW THAT GODDARD HADN'T ACTUALLY BUILT A WORKING MODEL YET.

THAT WOULD TAKE ANOTHER SEVEN YEARS OF TRIAL AND (MOSTLY) ERROR, THOUGH GODDARD WAS UNDETERRED.

WE LEARNED SOMETHING TODAY.

WE WON'T MAKE THE SAME MISTAKE AGAIN.

EVER THE OPTIMIST, HE REFERRED TO HIS FAILURES AS "NEGATIVE INFORMATION."

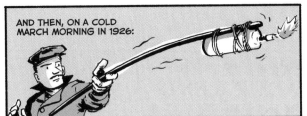

AND THEN, ON A COLD MARCH MORNING IN 1926:

FOOM

THE WORLD'S FIRST LIQUID-PROPELLANT ROCKET ROSE SKYWARD...

...AND CRASHED 2.5 SECONDS LATER.

FOR GODDARD, IT WAS NO LESS HISTORIC THAN THE WRIGHT BROTHERS' FIRST FLIGHT AT KITTY HAWK.

WHICH IS WHY IT WAS CURIOUS THAT HE URGED THE SMITHSONIAN NOT TO PUBLICIZE HIS SUCCESS.

"MY REASON IS THAT THIS ROCKET WORK IS BEING MADE ALMOST A NATIONAL ISSUE IN GERMANY."

CLAK CLAK CLAK CLAK CLAK CLAK CLAK CLAK

GODDARD WAS CONVINCED THAT HERMANN OBERTH WAS TRYING TO STEAL HIS IDEAS.

"NEARLY EVERY DAY, I AM IN RECEIPT OF REQUESTS FROM GERMANY FOR INFORMATION AND DETAILS."

INDEED, IT SEEMED THAT THE ONLY PEOPLE TAKING ROCKETRY SERIOUSLY WERE THE NATIONAL SOCIALISTS IN GERMANY.

THAT'S BECAUSE, AS PART OF THE TERMS OF THE TREATY OF VERSAILLES, WHICH ENDED WORLD WAR I, GERMANY HAD TO GUT ITS MILITARY.

THERE WAS NO MENTION IN THE TREATY, HOWEVER, OF ROCKETS.

IN 1932, BACK IN BERLIN, VON BRAUN WAS WORKING WITH OBERTH ON A TEST OF A NEW ROCKET EXPERIMENT WHEN AN UNFAMILIAR CAR APPEARED.

HERR VON BRAUN, I AM CAPTAIN WALTER DORNBERGER.

CAN YOU SPARE A MOMENT OF YOUR TIME FOR THE GERMAN ARMY?

A LITTLE MORE THAN A YEAR AFTER THIS CONVERSATION, ADOLF HITLER SEIZED CONTROL OF THE GERMAN GOVERNMENT.

ONE OF HITLER'S AMBITIONS FOR THE THIRD REICH WAS TO MAKE GERMANY A GREAT MILITARY POWER AGAIN.

THAT MEANT BUILDING NEW WEAPONS, MORE POWERFUL THAN ANY THE WORLD HAD EVER KNOWN.

AND FOR VON BRAUN AND HIS COLLEAGUES, THAT MEANT A SURGE OF FUNDING FOR THEIR ROCKETS.

BY THE TIME HE TURNED 25, WERNHER VON BRAUN WAS WELL ON HIS WAY TO REALIZING HIS LIFE'S DREAM.

77

IN SEPTEMBER 1935,

AS TSIOLKOVSKY WAS DYING OF CANCER

AND AS GODDARD WAS TOILING IN OBSCURITY IN NEW MEXICO,

HITLER ENACTED THE "NUREMBERG LAWS," WHICH REVOKED CITIZENSHIP FOR ANYONE OF JEWISH DESCENT AND DECLARED THEM "ENEMIES OF THE RACE-BASED STATE."

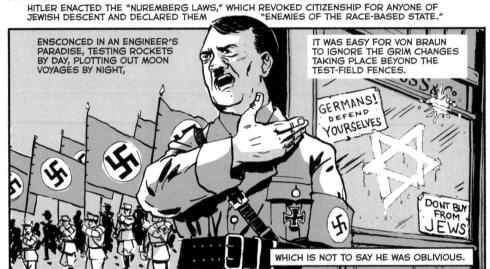

ENSCONCED IN AN ENGINEER'S PARADISE, TESTING ROCKETS BY DAY, PLOTTING OUT MOON VOYAGES BY NIGHT,

IT WAS EASY FOR VON BRAUN TO IGNORE THE GRIM CHANGES TAKING PLACE BEYOND THE TEST-FIELD FENCES.

GERMANS! DEFEND YOURSELVES

DON'T BUY FROM JEWS

WHICH IS NOT TO SAY HE WAS OBLIVIOUS.

HE, LIKE ALL HIS FELLOW ENGINEERS, HAD HAD TO PROVE HIS "ARYAN" ANCESTRY.

AS A STUDENT AT THE UNIVERSITY OF BERLIN,

HE MUST HAVE NOTICED THE PURGING OF JEWISH PROFESSORS AND STUDENTS.

FOR THE REST OF HIS LIFE, VON BRAUN WAS EVASIVE ABOUT HIS POLITICS:

MY FATHER WARNED ME THAT IT WAS ALL GOING TO END IN TRAGEDY FOR GERMANY AND FOR MANY OTHER PEOPLE TOO...

...BUT I WAS TOO WRAPPED UP IN ROCKETS TO HEED HIS WARNING.

NEVERTHELESS, IN THE WORDS OF HIS BIOGRAPHER:

"THERE CAN BE LITTLE DOUBT THAT HE WAS A LOYAL, PERHAPS EVEN MILDLY ENTHUSIASTIC SUBJECT OF HITLER'S DICTATORSHIP."

EVENTUALLY, VON BRAUN WOULD ACHIEVE THE RANK OF MAJOR IN THE SS.

ONE MORNING IN MARCH 1939, VON BRAUN AND HIS COLLEAGUES RECEIVED A VISITOR.

HITLER MADE A RARE JOURNEY TO PEENEMÜNDE, A SPRAWLING TEST FACILITY ON THE NORTH COAST OF GERMANY.

HE WAS THERE TO SEE VON BRAUN'S NEW ROCKET.

OFFICIALLY, IT WAS KNOWN AS "ASSEMBLY 4,"

BUT JOSEPH GOEBBELS, HITLER'S MINISTER OF PROPAGANDA, COINED A MORE DRAMATIC NAME:

"VENGEANCE WEAPON 2"

THE V-2.

EXPLAIN TO THE FÜHRER OUR PLANS FOR THE MISSILE, BUT DON'T WASTE HIS TIME WITH TALK ABOUT SPACEFLIGHT OR JOURNEYS TO THE MOON!

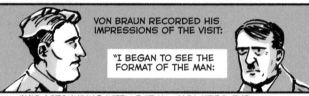

VON BRAUN RECORDED HIS IMPRESSIONS OF THE VISIT:

"I BEGAN TO SEE THE FORMAT OF THE MAN:

"HIS ASTOUNDING INTELLECTUAL CAPACITIES, THE ACTUALLY HYPNOTIC INFLUENCE OF HIS PERSONALITY...

"...HERE IS A NEW NAPOLEON, A NEW COLOSSUS, WHO HAS BROUGHT THE WORLD OUT OF EQUILIBRIUM."

HITLER SEEMED UNIMPRESSED.

PERHAPS, LIKE EVERYONE OUTSIDE OF VON BRAUN'S CIRCLE, HE DIDN'T TAKE ROCKETS VERY SERIOUSLY.

OR PERHAPS HE WAS SIMPLY DISTRACTED BY HIS PLANS FOR THE MONTHS AHEAD:

THE INVASION OF POLAND AND THE BEGINNING OF WORLD WAR II.

IN THE FIRST YEARS OF THE WAR, IT SEEMED AS IF HITLER'S TROOPS WERE UNSTOPPABLE.

WITH LIGHTNING SPEED, THE NAZIS CONQUERED MUCH OF EUROPE.

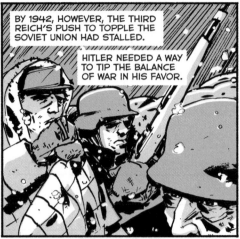

BY 1942, HOWEVER, THE THIRD REICH'S PUSH TO TOPPLE THE SOVIET UNION HAD STALLED.

HITLER NEEDED A WAY TO TIP THE BALANCE OF WAR IN HIS FAVOR.

MEANWHILE, IN PEENEMÜNDE:

3...2....

...1...

...IGNITION...

...MAIN STAGE.

LIFTOFF!

AT ITS APOGEE, VON BRAUN'S ROCKET SKIMMED THE UPPER REACHES OF THE MESOSPHERE, JUST SHY OF OUTER SPACE...

...BEFORE ARCING BACK EARTHWARD AT MORE THAN FOUR TIMES THE SPEED OF SOUND.

THAT EVENING, AT A CELEBRATION DINNER, DORNBERGER TOASTED THE DAY'S SUCCESS:

GENTLEMEN, THIS AFTERNOON THE SPACESHIP WAS BORN!

HEAR, HEAR!

LEST YOU THINK YOUR HEADACHES ARE OVER, I AM TELLING YOU, THEY ARE JUST BEGINNING.

IF WE ARE TO WIN THIS WAR, WE MUST LAUNCH, LAUNCH, LAUNCH.

HITLER EXPECTED THOUSANDS OF V-2'S, FAR MORE THAN THE SMALL GROUP AT PEENEMÜNDE COULD MANUFACTURE.

BUT SOON THEY WOULD HAVE THEIR SOLUTION: *SLAVERY.*

IN AUGUST 1943,

ON THE OUTSKIRTS OF A SMALL TOWN IN CENTRAL GERMANY,

DEEP IN THE BOWELS OF AN OLD GYPSUM MINE,

A NEW HELL WAS BORN.

IT WAS CALLED *MITTELWERK*, A SECRET UNDERGROUND FACTORY FOR VON BRAUN'S ROCKETS.

HANS KAMMLER, THE SS OFFICER WHO DESIGNED THE GAS CHAMBERS AT AUSCHWITZ, WAS THE MAN IN CHARGE.

PAY NO ATTENTION TO THE HUMAN COST.

THE WORK MUST GO AHEAD, AND IN THE SHORTEST POSSIBLE TIME.

BY THE TIME VON BRAUN VISITED THE FACTORY IN NOVEMBER,

MORE THAN 10,000 CONCENTRATION CAMP PRISONERS LIVED AND WORKED IN THE TUNNELS.

THE FACTORY AT MITTELWERK WAS ESSENTIALLY A DEATH CAMP.

WITHOUT RUNNING WATER OR HEAT,

CHOKING ON FUMES AND SMOKE,

SLOWLY STARVING AND WEAKENED BY DISEASE,

THE PRISONERS WORKED FIFTEEN-HOUR SHIFTS IN THE NEAR DARK.

BY DECEMBER, PRISONERS WERE DYING AT THE RATE OF 20 A DAY.

YEARS LATER, VON BRAUN RECALLED A VISIT TO THE FACTORY:

IT IS REPULSIVE TO BE SUDDENLY SURROUNDED BY PRISONERS.

THE WHOLE ATMOSPHERE WAS UNBEARABLE.

AS REPULSED AS HE MAY HAVE BEEN, VON BRAUN REFUSED TO ABANDON HIS DREAM. HIS BARGAIN WITH THE NAZIS HAD COME DUE.

WAR IS WAR, AND BECAUSE MY COUNTRY FOUND ITSELF AT WAR, I HAD THE CONVICTION THAT I DID NOT HAVE THE RIGHT TO BRING FURTHER MORAL VIEWPOINTS TO BEAR.

FOR THEIR PART, THE PRISONERS RESISTED AS MUCH AS THEY COULD.

SNIP

IN RESPONSE, HANS KAMMLER ORDERED DAILY MASS HANGINGS.

THE CAMP CREMATORIUM WAS RARELY COLD.

THE FIRST 50 ROCKETS TO COME OUT OF MITTELWERK WERE UNUSABLE.

NEVERTHELESS, THE FACTORY EXPANDED.

TENS OF THOUSANDS OF CAPTIVES ROLLED IN ON CRAMPED BOXCARS.

SOON, HITLER WOULD HAVE HIS WONDER WEAPON.

IN LATE 1944, THE FIRST V-2'S SCREAMED OVER THE SKIES OF PARIS AND LONDON.

THOUSANDS MORE WOULD FOLLOW.

IN TOTAL, THE NAZIS LAUNCHED MORE THAN 3000 V-2 ROCKETS,

MOST OF THEM TARGETING LONDON AND ANTWERP.

SOME 5000 PEOPLE WERE KILLED IN THE BARRAGE AND MANY MORE THOUSANDS INJURED.

HARROWING STATISTICS, BUT WHEN SEEN IN THE LARGER CONTEXT OF THE WAR--

IN WHICH THE U.S. BOMBING RAID OF TOKYO INCINERATED 100,000 PEOPLE OVERNIGHT--

THE V-2 ROCKET WAS A NOVEL, IF NOT PARTICULARLY DEADLY, WEAPON.

ACTUALLY, IN ONE WAY, IT *WAS* QUITE DEADLY:

FOR EVERY ONE PERSON KILLED BY V-2 STRIKES, TWO PRISONERS DIED ON THE ASSEMBLY LINE AT MITTELWERK.

BY THE END OF 1944, IT WAS CLEAR THAT THE V-2 WOULD NOT CHANGE THE COURSE OF THE WAR.

AND WERNHER VON BRAUN HAD LOST WHATEVER ZEAL HE ONCE HAD FOR THE NAZIS.

ONE NIGHT IN MARCH, THE GESTAPO PAID HIM A VISIT.

BANG BANG

AM I UNDER ARREST? THERE MUST BE A MISUNDERSTANDING!

WE HAVE ORDERS TO TAKE YOU UNDER "PROTECTIVE" CUSTODY.

DO YOU DENY THE CHARGE OF SABOTAGING THE REICH'S ROCKET PROGRAM TO FUND YOUR OBSESSION WITH SPACEFLIGHT?

DO YOU DENY YOUR INTENTION TO FLEE TO ENGLAND WITH THE V-2 PLANS?

THERE WAS NO PROOF, HOWEVER, AND THE CHARGES WERE SOON DROPPED.

VON BRAUN WENT BACK TO WORK, BUT HE WAS SHAKEN BY THE EPISODE.

I HAVE ALREADY PACKED MY TRUNKS WITH THE DOCUMENTATION, AND I WILL OFFER MY SERVICES TO THE AMERICANS.

THEN I WILL BUILD MY SPACE ROCKET.

IN APRIL 1945, WHEN PRIVATE GALIONE AND THE 104TH INFANTRY FOUND MITTELWERK,

THEY STUMBLED UPON A TREASURE TROVE OF INFORMATION.

ALL TOLD, THE U.S. ACQUIRED 14 TONS OF MATÉRIEL--PARTS, PLANS, EVEN SEVERAL FULLY ASSEMBLED ROCKETS.

WHAT THEY COULDN'T REMOVE, THEY SIMPLY DESTROYED.

WITH THE SOVIET ARMY APPROACHING FROM THE NORTH, THE AMERICANS WERE RACING AGAINST TIME.

GERMANY

BOTH SIDES WERE EAGER TO PLUNDER NAZI MILITARY TECHNOLOGY.

BY THE TIME THE RUSSIANS ARRIVED, THE AMERICANS WERE GONE AND THE FACTORY WAS EMPTY.

AMONG THE DETRITUS WAS A WORN COPY OF TSIOLKOVSKY'S PAPER.

IN THE MARGINS WERE VON BRAUN'S NOTES.

JOSEF STALIN WAS LIVID WHEN HE HEARD THE NEWS.

THIS IS ABSOLUTELY INTOLERABLE.

WE DEFEATED THE NAZI ARMIES, WE OCCUPIED BERLIN...

BUT THE AMERICANS GOT THE ROCKET ENGINEERS.

INTELLIGENCE AGENTS, MEMBERS OF THE OFFICE OF STRATEGIC SERVICES (THE PRECURSOR TO THE CIA), WERE SCOURING EUROPE,

TRACKING DOWN GERMAN SCIENTISTS IN AN OPERATION CODE-NAMED PAPERCLIP.

WHEN PRESIDENT TRUMAN AUTHORIZED OPERATION PAPERCLIP, HE EXPLICITLY REFUSED TO GRANT AMERICAN CITIZENSHIP TO FORMER NAZI OFFICERS.

WHAT TRUMAN DIDN'T KNOW WAS THAT OSS OPERATIVES WERE SECRETLY FORGING PAPERS FOR THE SCIENTISTS, COVERING UP ANY INCONVENIENT TRUTHS.

SO BEGAN THE WHOLESALE EXODUS OF NAZI SCIENTISTS TO THE UNITED STATES.

IN LESS THAN A YEAR, WERNHER VON BRAUN HAD TRAVELED FROM THE TUNNELS OF MITTELWERK TO THE HALLS OF THE PENTAGON.

HE WAS IMMEDIATELY PUT TO WORK BUILDING BIGGER, BETTER VERSIONS OF THE V-2 ROCKET.

HE ALSO PLAYED THE UNLIKELY PART OF PUBLIC AMBASSADOR, A SALESMAN FOR THE CAUSE OF SPACE TRAVEL.

IN 1955, WALT DISNEY FEATURED HIM IN AN ANIMATED TELEVISION SERIES, *MAN IN SPACE*.

THAT SAME YEAR, HE AND 40 OTHER GERMAN SCIENTISTS BECAME AMERICAN CITIZENS.

THE HANDSOME, CONFIDENT MAN WITH THE ARISTOCRATIC PRUSSIAN ACCENT WAS, FOR MILLIONS OF SCHOOLCHILDREN, THE PUBLIC FACE OF THE AMERICAN SPACE PROGRAM.

THE PRESS DOWNPLAYED HIS NAZI PAST.

HIS BIOGRAPHY IN ONE MAGAZINE SIMPLY READ:

"AT FORTY, HE IS CONSIDERED THE FOREMOST ROCKET ENGINEER IN THE WORLD TODAY."

"HE WAS BROUGHT TO THIS COUNTRY FROM GERMANY BY THE U.S. GOVERNMENT IN 1945."

THE HISTORIAN MICHAEL NEUFELD SUMMARIZED THE DIZZYING TRAJECTORY OF WERNHER VON BRAUN'S CAREER:

"HERE WAS A MAN WHO HAD SHAKEN THE HANDS OF EISENHOWER, KENNEDY, JOHNSON, AND NIXON...

"BUT ALSO HITLER, HIMMLER, GÖRING, AND GOEBBELS."

WHICH MADE HIM A UNIQUE PROPHET FOR AMERICA'S FUTURE.

WITHIN THE NEXT 10 OR 15 YEARS, THE EARTH WILL HAVE A NEW COMPANION IN THE SKIES,

A SATELLITE THAT COULD EITHER BE THE GREATEST FORCE FOR PEACE EVER DEVISED

OR ONE OF THE MOST TERRIBLE WEAPONS OF WAR-- DEPENDING ON WHO MAKES AND CONTROLS IT.

WERNHER VON BRAUN COULD ALREADY SEE THE TERRIFYING AND TRANSCENDENT FUTURE THAT HIS ROCKETS MADE POSSIBLE.

HE WOULD BUILD MORE OF THEM FOR HIS NEW HOMELAND, BUILD THEM BIGGER AND FARTHER-REACHING.

AND HE TOLD HIMSELF HE WOULD NEVER AGAIN BE ON THE LOSING SIDE.

CHAPTER IX
EAGLE

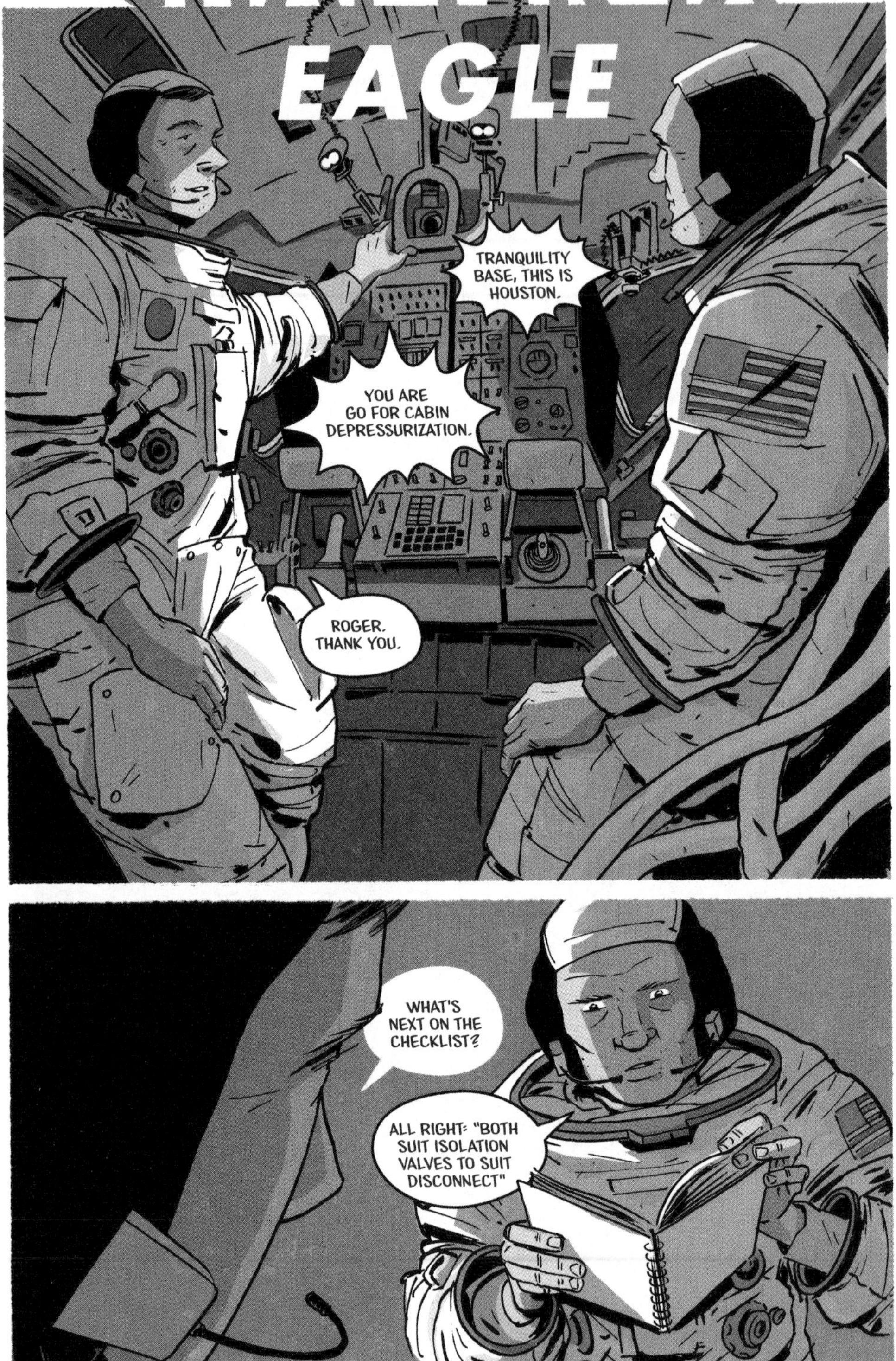
TRANQUILITY BASE, THIS IS HOUSTON.
YOU ARE GO FOR CABIN DEPRESSURIZATION.
ROGER. THANK YOU.
WHAT'S NEXT ON THE CHECKLIST?
ALL RIGHT: "BOTH SUIT ISOLATION VALVES TO SUIT DISCONNECT"

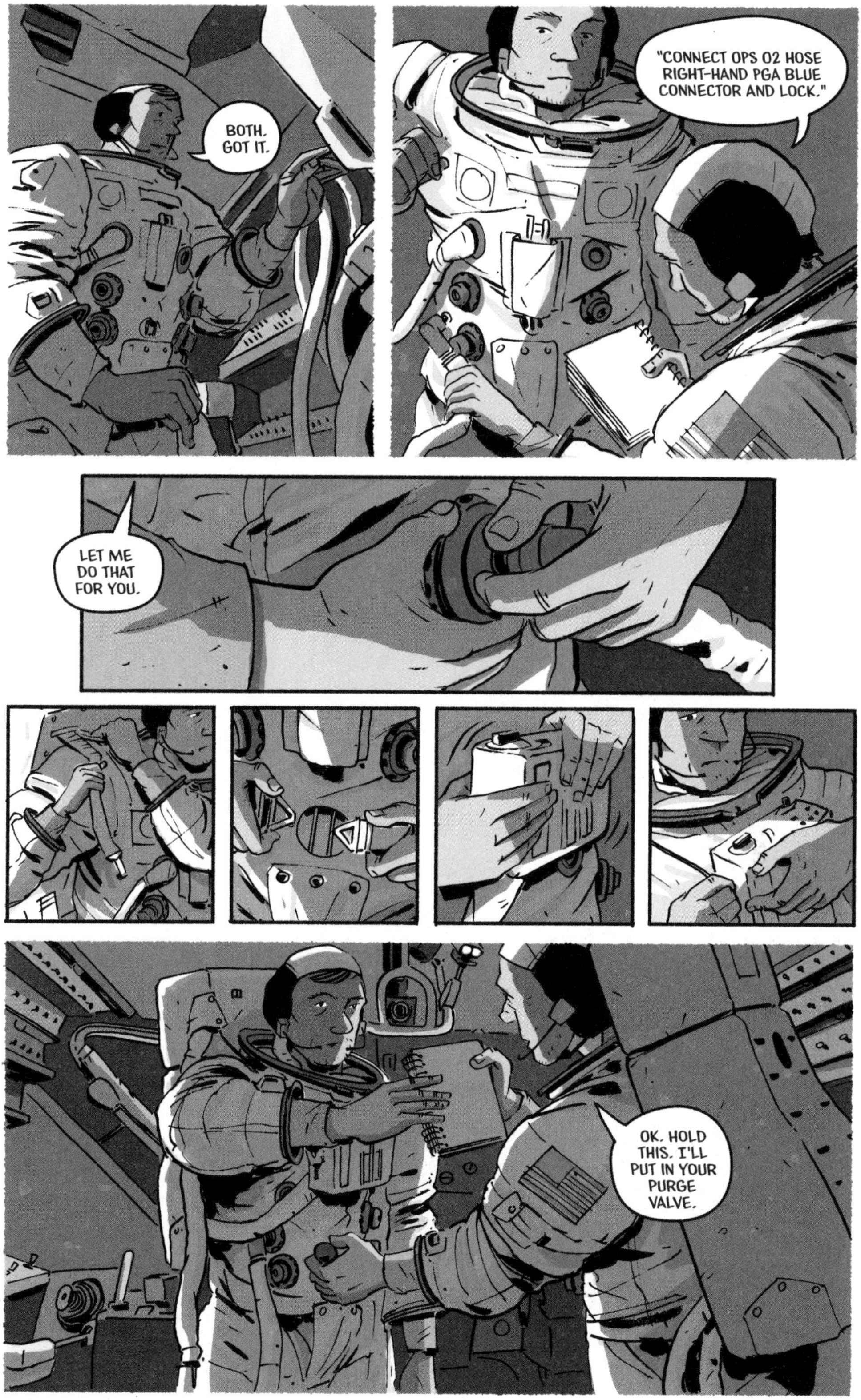
BOTH. GOT IT.
"CONNECT OPS O2 HOSE RIGHT-HAND PGA BLUE CONNECTOR AND LOCK."
LET ME DO THAT FOR YOU.
OK. HOLD THIS. I'LL PUT IN YOUR PURGE VALVE.

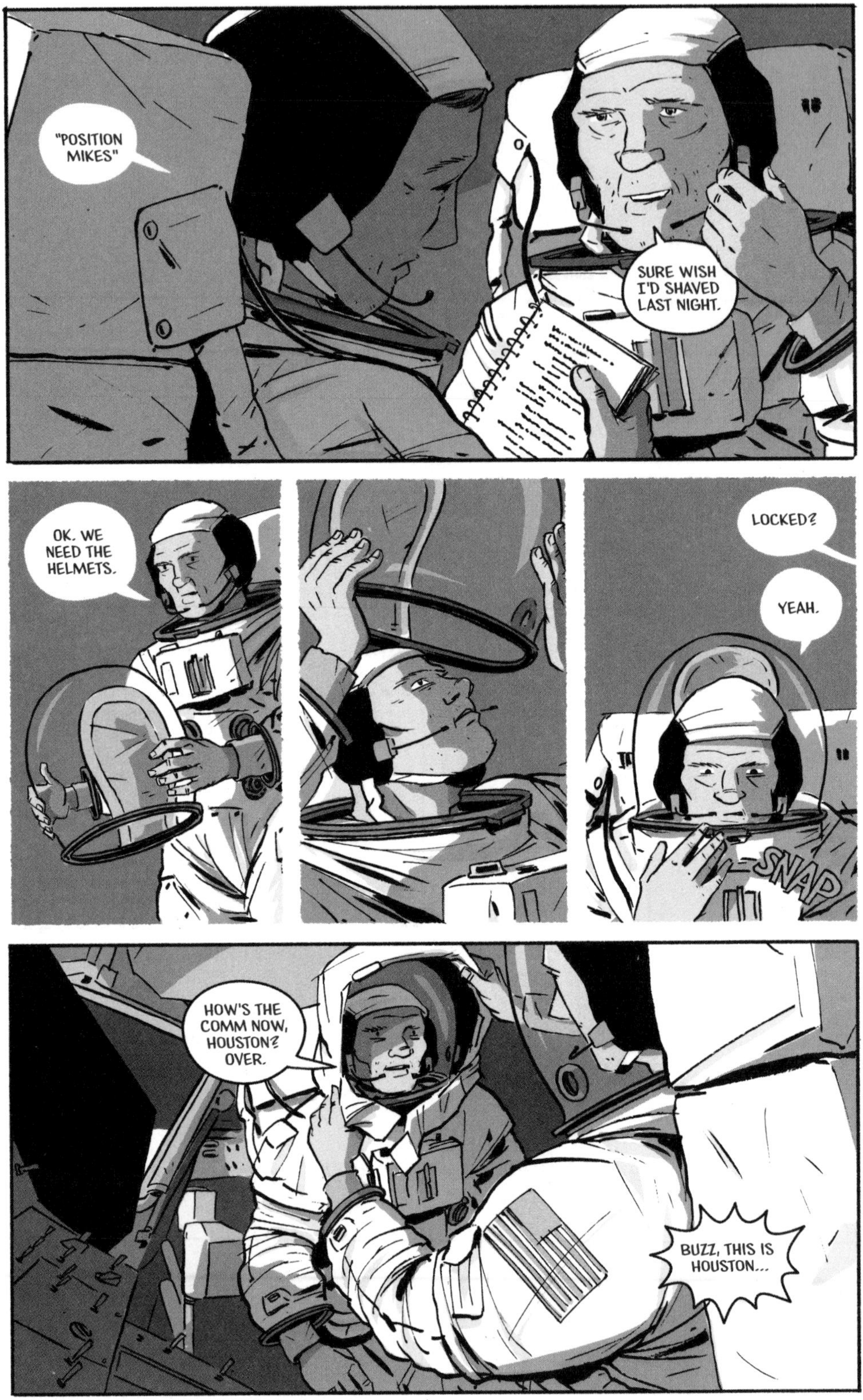
"POSITION MIKES"
SURE WISH I'D SHAVED LAST NIGHT.
OK. WE NEED THE HELMETS.
LOCKED?
YEAH.
SNAP
HOW'S THE COMM NOW, HOUSTON? OVER.
BUZZ, THIS IS HOUSTON...

...YOU'RE COMING IN LOUD AND CLEAR.

VERY WELL. THANK YOU.

OK. WE CAN STOW THIS.

ALL RIGHT. NOW COMES THE GYMNASTICS.

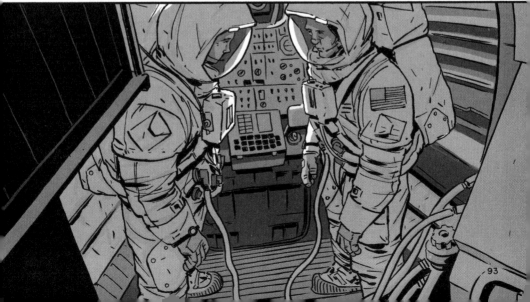

CHAPTER X
SATELLITES

WHOSE IMAGINATION IS NOT FIRED BY THE POSSIBILITY
OF VOYAGING OUT BEYOND THE LIMITS OF OUR EARTH,
TRAVELING TO THE MOON, TO VENUS AND MARS? SUCH
THOUGHTS WHEN PUT ON PAPER NOW SEEM LIKE IDLE
FANCY. BUT, A MAN-MADE SATELLITE, CIRCLING OUR
GLOBE BEYOND THE LIMITS OF THE ATMOSPHERE IS THE
FIRST STEP.

--RAND CORPORATION

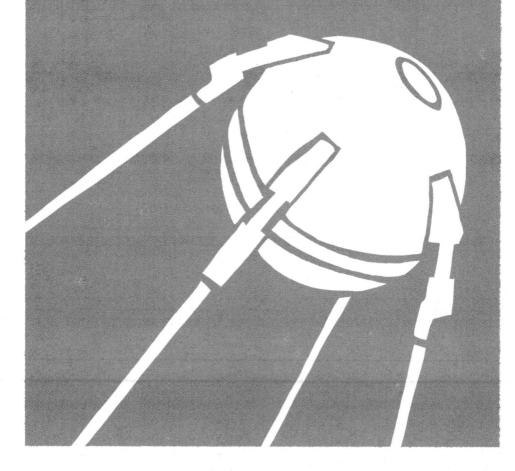

THE RUSSIANS CALLED IT WHAT IT WAS:

"A FELLOW TRAVELER OF THE EARTH."

SPUTNIK ZEMLYI.

IT WAS BUILT TO BE AS SIMPLE,

AS COMPACT,

AND AS LIGHTWEIGHT AS POSSIBLE.

THE SATELLITE HAD ONE JOB TO DO:

TO BE THE FIRST HUMAN-MADE THING

TO REACH EARTH ORBIT.

SPUTNIK WAS THE BRAINCHILD OF A GROUP OF RUSSIAN VISIONARIES,

LED BY A FORMER CONVICT

WHOSE VERY NAME WAS A STATE SECRET.

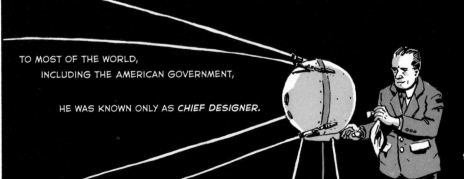

TO MOST OF THE WORLD,

INCLUDING THE AMERICAN GOVERNMENT,

HE WAS KNOWN ONLY AS *CHIEF DESIGNER.*

95

MAY 1940.

SERGEI PAVLOVICH KOROLEV SHOULD HAVE BEEN DEAD.

HE SHOULD HAVE DIED IN 1938,

WHEN THE SOVIET SECRET POLICE CAPTURED AND TORTURED HIM DURING THE GREAT PURGE.

PLEASE, LET ME SAY GOODBYE TO MY FAMILY!

HE SHOULD HAVE DIED AT THE HANDS OF ONE OF STALIN'S EXECUTIONERS,

LIKE THE MILLIONS OF OTHERS FORCED TO CONFESS TO TRUMPED-UP CHARGES.

TODAY IS YOUR TRIAL. YOU ARE ACCUSED OF SABOTAGE. CONFESS!

I DID NOT... COMMIT...A CRIME.

THAT'S WHAT EVERY CRIMINAL SAYS.

96

AND HE SHOULD HAVE DIED SERVING HIS SENTENCE IN A SIBERIAN LABOR CAMP,

WHERE ONE IN FOUR OF HIS FELLOW CONVICTS FROZE, OR CAUGHT SICK, OR FELL AFOUL OF THE GUARDS.

"WHEN I OPENED MY EYES, I COULD SEE SOMETHING FLUTTERING...

"...A BUTTERFLY, SOMETHING ON THIS EARTH STILL ALIVE AND BEAUTIFUL.

"I WAS ALIVE."

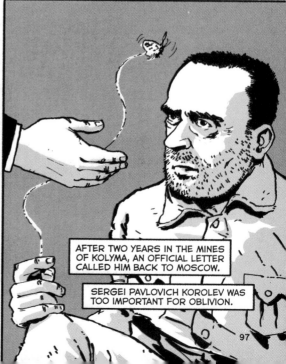

AFTER TWO YEARS IN THE MINES OF KOLYMA, AN OFFICIAL LETTER CALLED HIM BACK TO MOSCOW.

SERGEI PAVLOVICH KOROLEV WAS TOO IMPORTANT FOR OBLIVION.

97

BEFORE HIS ARREST, KOROLEV HAD BEEN A PROMISING YOUNG ENGINEER,

A DESIGNER OF ROCKET-POWERED AIRPLANES.

THE GREAT PURGE HAD PUT MANY OF THE SOVIET UNION'S BEST MINDS IN PRISON.

BUT BY 1941, STALIN FACED A BIGGER THREAT THAN THE DEMONS CONJURED BY HIS PARANOIA.

THAT SUMMER, HITLER INVADED THE SOVIET UNION.

IF THE NATION WAS TO SURVIVE, IT NEEDED MORE POWERFUL WEAPONS.

IN MOSCOW, KOROLEV WAS STILL A PRISONER, BUT INSTEAD OF HARD LABOR, HIS SENTENCE WAS TO GO BACK TO CRAFTING AIRPLANE ENGINES.

WELCOME TO YOUR NEW HOME, CRIMINAL.

OUR COUNTRY DOESN'T NEED YOUR FIREWORKS.

OR MAYBE YOUR ROCKETS ARE FOR AN ATTEMPT ON THE LIFE OF OUR LEADER?

AT THE TIME, THE SOVIETS HAD NO USE FOR SPACE ROCKETS.

BUT BY THE WAR'S END, TWO REVELATIONS BEGAN TO CHANGE STALIN'S MIND ABOUT ROCKETS:

FIRST, THE RED ARMY CAPTURED PEENEMÜNDE AND MITTELWERK, EXPOSING THE FULL EXTENT OF THE GERMAN V-2 PROGRAM.

AND THEN, ON AUGUST 6, 1945, THE UNITED STATES OBLITERATED THE JAPANESE CITY OF HIROSHIMA WITH AN APOCALYPTIC NEW WEAPON...

THE ATOMIC BOMB.

WITH AN EXPLOSIVE YIELD THAT WAS ORDERS OF MAGNITUDE GREATER THAN ANYTHING IN THE HISTORY OF WARFARE,

A SINGLE ATOMIC BOMB COULD VAPORIZE A CITY IN AN INSTANT,

LEAVING ONLY A VOID OF RUIN AND TOXIC RADIOACTIVE FALLOUT.

IN THE WORDS OF ONE SURVIVOR:

"SUCH A WEAPON HAS THE POWER TO MAKE EVERYTHING INTO NOTHING."

ASIDE FROM ITS CAPACITY FOR DESTRUCTION,

THE ATOMIC BOMB CHANGED THE VERY DEFINITION OF WHAT IT MEANT TO BE A POWERFUL NATION.

HIROSHIMA HAS SHAKEN THE WHOLE WORLD. THE BALANCE HAS BEEN DESTROYED.

STALIN RECOGNIZED THAT WITHOUT AN ATOMIC WEAPON OF ITS OWN,

THE U.S.S.R. WOULD HAVE LITTLE INFLUENCE OVER GLOBAL POLITICS.

BUT A NUCLEAR DEVICE ALONE WAS USELESS WITHOUT A WAY TO DEPLOY IT.

STALIN NEEDED MORE THAN JUST A BOMB.

HE NEEDED A MISSILE THAT COULD REACH THE UNITED STATES.

THE YEARS AFTER THE WAR WERE BUSY FOR SERGEI KOROLEV. HE WAS SENT TO GERMANY TO SALVAGE WHATEVER WAS LEFT OF THE V-2 PROGRAM...

WE KNOW PRACTICALLY NOTHING ABOUT THE MISSILE EXCEPT THAT IT FLIES.

...AND TO RECRUIT THE GERMAN ROCKET EXPERTS WHO HADN'T FOLLOWED VON BRAUN TO AMERICA.

IF WE INCREASE THE RANGE, WE WILL FINALLY BE ABLE TO BUILD ARTIFICIAL SATELLITES.

AND IF WE INCREASED THE CUTOFF VELOCITY BY ABOUT...SAY 40 PERCENT...

...THEN WE COULD VISIT THE MOON.

LET'S ALL WORK TOGETHER TO ACHIEVE THIS.

THEIR IMMEDIATE CHALLENGE WAS TO REVERSE-ENGINEER THE V-2.

FROM THERE, THEY COULD BUILD A BETTER VERSION, WITH A LONGER RANGE, THAT COULD CARRY A HEAVIER BOMB.

THE AMERICANS WERE LIKELY ALREADY DOING THE SAME THING, BUT WITH A HEAD START.

AT THE WHITE SANDS PROVING GROUND, IN NEW MEXICO--

ONLY A FEW MILES FROM WHERE THE FIRST ATOMIC BOMB WAS TESTED--

WERNHER VON BRAUN AND HIS TEAM WERE IMPROVING ON THEIR V-2 DESIGNS.

WHILE FARTHER WEST, IN SANTA MONICA, CALIFORNIA,

AN ECLECTIC GROUP OF SPECIALISTS WAS BRAINSTORMING ABOUT HOW BEST TO APPLY VON BRAUN'S INNOVATIONS.

THOUGH THE CRYSTAL BALL IS CLOUDY, TWO THINGS SEEM CLEAR:

A SATELLITE VEHICLE...CAN BE EXPECTED TO BE ONE OF THE MOST POTENT SCIENTIFIC TOOLS OF THE 20TH CENTURY.

THEY WERE MEMBERS OF WHAT WOULD BECOME THE RAND CORPORATION, THE WORLD'S FIRST THINK TANK,

THE ACHIEVEMENT OF A SATELLITE CRAFT BY THE UNITED STATES WOULD INFLAME THE IMAGINATION OF MANKIND.

AND IN 1946 THEY WERE COMMISSIONED BY THE ARMY AIR FORCE TO RESEARCH THE POTENTIAL SIGNIFICANCE OF WHAT THEY CALLED AN "EARTH-CIRCLING SPACESHIP."

RIGHT, AND IT WOULD PROBABLY PRODUCE REPERCUSSIONS IN THE WORLD COMPARABLE TO THE EXPLOSION OF THE ATOMIC BOMB.

THE RAND REPORT WAS A DECADE AHEAD OF ITS TIME, BUT IT SET AN IMPORTANT PRECEDENT FOR AMERICA'S ROLE IN THE ATOMIC AGE:

THAT A NATION'S SENSE OF POWER AND SECURITY WAS TIED TO ITS CAPACITY FOR TECHNOLOGICAL INNOVATION...

...AND ITS ABILITY TO EXPLOIT THE NEW FRONTIER OF SPACE.

THE RAND REPORT GOT A LOT OF THINGS RIGHT ABOUT THE COMING SPACE AGE,

HMMPFF...

BUT IN 1946, AN "EARTH-CIRCLING SPACESHIP" STILL SEEMED LIKE THE STUFF OF SCIENCE FICTION.

PRESIDENT HARRY TRUMAN WAS MORE PRAGMATIC.

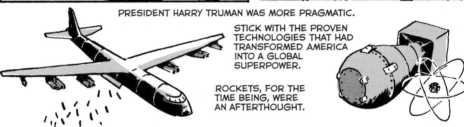

STICK WITH THE PROVEN TECHNOLOGIES THAT HAD TRANSFORMED AMERICA INTO A GLOBAL SUPERPOWER.

ROCKETS, FOR THE TIME BEING, WERE AN AFTERTHOUGHT.

MEANWHILE, THE SOVIETS WERE SIMPLY TRYING TO FIGURE OUT WHERE TO START.

THE COUNTRY WAS DEEPLY SCARRED BY THE WAR.

THE WESTERN CITIES WERE RUINS; SOME 20 MILLION SOLDIERS AND CITIZENS WERE DEAD.

AND COMPARED WITH THE BRITISH AND AMERICANS, THE U.S.S.R. WAS TECHNOLOGICALLY BACKWARD.

I WAS TOLD TO REPORT TO MR. KOROLEV. IS THIS THE RIGHT PLACE?

YEAH YEAH, HE'S ON HIS WAY.

HEY, COULD YOU SLIDE THAT BUCKET OVER A LITTLE?

THERE WAS A LOT OF CATCHING-UP TO DO.

RECOGNIZING KOROLEV'S PROGRESS WITH GERMAN TECHNOLOGY, STALIN MADE HIM CHIEF DESIGNER OF ALL LONG-RANGE ROCKETS.

FOR A MAN WHO WAS STILL TECHNICALLY AN ENEMY OF THE STATE, THIS WAS AN INCREDIBLE TURN OF EVENTS.

IT'S HARD TO IMAGINE WHAT KOROLEV MUST HAVE FELT WHEN HE FIRST SHOOK THE HAND OF THE MAN WHOSE PURGES CONDEMNED HIM TO YEARS OF FALSE IMPRISONMENT.

KOROLEV RARELY TALKED ABOUT SUCH PERSONAL FEELINGS.

AS ONE OF HIS COLLEAGUES PUT IT:

WE DEVELOPED A KIND OF VOLUNTARY SCHIZOPHRENIA IN THOSE DAYS.

YOU SEPARATE YOUR SOUL INTO TWO INDEPENDENT PARTS.

YOU CAN ADAPT YOURSELF TO THE PRACTICE OF LYING IN ONE PART OF YOUR SOUL WHILE YOU KEEP THE TRUTH IN THE OTHER PART.

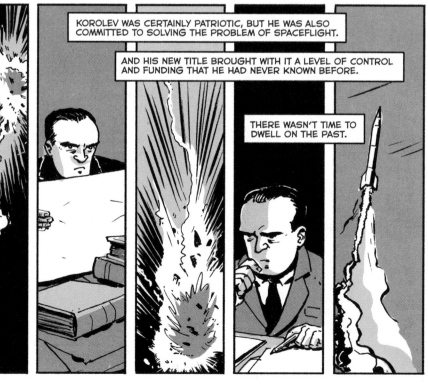

KOROLEV WAS CERTAINLY PATRIOTIC, BUT HE WAS ALSO COMMITTED TO SOLVING THE PROBLEM OF SPACEFLIGHT.

AND HIS NEW TITLE BROUGHT WITH IT A LEVEL OF CONTROL AND FUNDING THAT HE HAD NEVER KNOWN BEFORE.

THERE WASN'T TIME TO DWELL ON THE PAST.

MEANWHILE, IN 1948,

A QUIRKY SOVIET ENGINEER NAMED MIKHAIL TIKHONRAVOV HAD COME UP WITH A RADICAL IDEA FOR A SPACESHIP DESIGN.

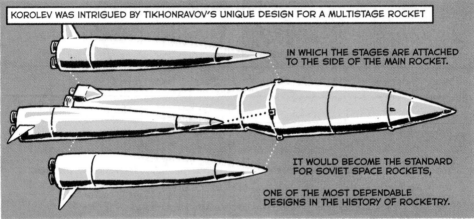

KOROLEV WAS INTRIGUED BY TIKHONRAVOV'S UNIQUE DESIGN FOR A MULTISTAGE ROCKET

IN WHICH THE STAGES ARE ATTACHED TO THE SIDE OF THE MAIN ROCKET.

IT WOULD BECOME THE STANDARD FOR SOVIET SPACE ROCKETS,

ONE OF THE MOST DEPENDABLE DESIGNS IN THE HISTORY OF ROCKETRY.

WHAT EXCITED HIM MORE, HOWEVER, WAS WHERE TIKHONRAVOV WANTED TO GO WITH THESE ROCKETS.

IT DOES NOT SEEM UNREASONABLE TO THINK THAT WITHIN A DECADE WE COULD SEND HUMANS INTO SPACE, OR EVEN TO THE MOON.

KOROLEV AGREED, BUT WAS MORE DISCREET WHEN TALKING ABOUT SPACEFLIGHT.

BY THE PARANOID LOGIC OF THE SECRET POLICE, ANY WORK THAT DIDN'T HAVE AN EXPLICIT MILITARY USE WAS TREASON.

BUT MAYBE THERE WAS A WAY TO DO BOTH:

TO BUILD STALIN THE WORLD'S MOST POWERFUL ROCKET

AND TO USE IT AS BOTH A SPACESHIP *AND* A WEAPON.

NO ONE WHO WORKED ON THE MANHATTAN PROJECT--

THE AMERICAN EFFORT TO DESIGN AN ATOMIC BOMB--

EXPECTED THAT THE SECRET TO NUCLEAR WEAPONS WOULD STAY SECRET FOR VERY LONG.

BUT SURELY IT WOULD TAKE LONGER THAN A FEW YEARS.

WHICH IS WHY PRESIDENT HARRY TRUMAN'S ANNOUNCEMENT IN 1949 CAME AS SUCH A SHOCK:

"WE HAVE EVIDENCE THAT WITHIN RECENT WEEKS AN ATOMIC EXPLOSION OCCURRED IN THE U.S.S.R."

AMERICA HAD LOST ITS NUCLEAR MONOPOLY.

AT FIRST, THE AMERICAN RESPONSE WAS TO BUILD BIGGER WEAPONS:

THE HYDROGEN BOMB.

"IVY MIKE": 10 MEGATONS

NAGASAKI: 19 KILOTONS

HIROSHIMA: 16 KILOTONS

BUT THE SOVIETS WERE NEVER FAR BEHIND.

WHILE IT WAS CLEAR THAT NATIONAL SECURITY IN THE ATOMIC AGE HINGED ON A GOVERNMENT'S ABILITY TO DEVELOP THE MOST ADVANCED WEAPONS TECHNOLOGY,

EVENTUALLY THE NUCLEAR ARSENALS BECAME SO DESTRUCTIVE THAT IF THE UNITED STATES AND THE SOVIET UNION ACTUALLY USED THEM AGAINST EACH OTHER,

THE RESULT MIGHT VERY WELL BE THE END OF CIVILIZATION.

SINCE ALL-OUT WAR WAS SUICIDAL, THE NEW BATTLEFRONT WOULD HAVE TO BE A DIFFERENT KIND OF FIGHT:

A "COLD" WAR, WAGED WITH POLITICAL AND IDEOLOGICAL WEAPONS.

IN THE EARLY 1950'S, IT SEEMED AS IF A SATELLITE COULD BE USED AS JUST SUCH A WEAPON.

SEVERAL SECRET REPORTS, FROM THE RAND CORPORATION,

THE DEPARTMENT OF DEFENSE,

A SMALL SCIENTIFIC SATELLITE COULD BE DEVELOPED FROM EXISTING MISSILE PROGRAMS ALREADY UNDER WAY.

AND FROM WITHIN THE EISENHOWER ADMINISTRATION ALL CAME TO SIMILAR CONCLUSIONS:

IF THE SOVIET UNION SHOULD ACCOMPLISH THIS AHEAD OF US, IT WOULD BE A SERIOUS BLOW TO THE PRESTIGE OF AMERICA.

THE SUCCESSFUL LAUNCHING OF A SATELLITE IS BOUND TO CAUSE A WORLDWIDE SENSATION.

WERNHER VON BRAUN, EVER IMPATIENT TO GET TO SPACE, WROTE HIS OWN APPEAL IN 1954.

BY LIMITING THE PAYLOAD TO FIVE POUNDS, MY MINIMUM SATELLITE VEHICLE COULD BE PUT INTO ORBIT WITHIN A FEW YEARS, USING ALREADY EXISTING ROCKET TECHNOLOGY.

IT WOULD BE SMALL, CHEAP, QUICK TO BUILD, SIMPLE, AND ABOVE ALL:

IT WOULD BE THE FIRST.

A MAN-MADE SATELLITE, NO MATTER HOW HUMBLE, WOULD BE A SCIENTIFIC ACHIEVEMENT OF TREMENDOUS IMPACT.

BUT WHAT FINALLY CONVINCED EISENHOWER TO FUND A SATELLITE PROGRAM WAS A CURIOSITY OF A DIFFERENT KIND...

106

EISENHOWER WANTED TO KNOW WHAT THE SOVIET MILITARY WAS UP TO.

AT THE TIME, EVEN BASIC INTELLIGENCE WAS HARD TO COME BY.

THE SOVIET LEADERSHIP, STEEPED IN THE PARANOIA OF STALIN'S RULE, GUARDED ITS WEAPONS PROGRAMS WITH FANATICAL SECRECY.

THAT SECRECY WOULD BE NO MATCH FOR A SATELLITE WITH A CAMERA.

TOO HIGH UP TO SHOOT DOWN, AN ORBITING EYE IN THE SKY COULD SURVEY THE ENTIRE SOVIET UNION WITH IMPUNITY. OF COURSE, IF THE SOVIETS EVER FOUND OUT, THEY MIGHT CONSIDER IT AN ACT OF WAR.

ACTUALLY, THIS RAISED AN INTERESTING QUESTION: HOW HIGH UP DOES A NATION'S SOVEREIGNTY EXTEND?

SURELY A COUNTRY COULDN'T CLAIM TO OWN THE INFINITUDE OF SPACE ABOVE ITS BORDERS...

...BUT IN 1954 THE UNITED NATIONS HAD YET TO SETTLE ON AN ANSWER TO THIS QUESTION.

EISENHOWER'S ADVISERS SUGGESTED THAT A CIVILIAN SATELLITE, ON A SCIENTIFIC MISSION,

PRELIMINARY STUDIES INDICATE THAT THERE IS NO OBSTACLE UNDER INTERNATIONAL LAW TO THE LAUNCHING OF SUCH A SATELLITE.

ORBITING OVER THE BORDERS OF DOZENS OF COUNTRIES,

WOULD SET AS A LEGAL PRECEDENT THE "FREEDOM OF SPACE" FOR ALL NATIONS.

IT WAS A GRAND GESTURE IN SERVICE OF A SINGLE, SECRETIVE GOAL:

THE UNRESTRICTED MOVEMENT OF AMERICAN SPY SATELLITES.

107

THE ONLY PROBLEM WITH THIS PLAN FOR A CIVILIAN SATELLITE WAS THAT THE ROCKETS FOR LAUNCHING IT WERE ALL MILITARY PROJECTS.

EISENHOWER TOOK A DIFFERENT ROUTE: PROJECT VANGUARD.

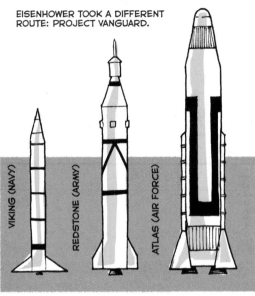

VIKING (NAVY)

REDSTONE (ARMY)

ATLAS (AIR FORCE)

INSTEAD OF A MISSILE, VANGUARD WOULD BE A "SOUNDING" ROCKET--A TOOL USED FOR GATHERING METEOROLOGICAL DATA.

AND WITH ROOM FOR LITTLE MORE THAN A SMALL, GRAPEFRUIT-SIZED SATELLITE,

THERE WOULD BE NO CONFUSION ABOUT PURPOSE: THIS WAS NO WEAPON.

JULY 25, 1955

THE PRESIDENT HAS APPROVED PLANS FOR GOING AHEAD WITH THE LAUNCHING OF SMALL UNMANNED EARTH-CIRCLING SATELLITES...

...AS PART OF THE U.S. PARTICIPATION IN THE INTERNATIONAL GEOPHYSICAL YEAR, WHICH TAKES PLACE BETWEEN JULY 1957 AND DECEMBER 1958.

THIS PROGRAM WILL FOR THE FIRST TIME IN HISTORY

ENABLE SCIENTISTS THROUGHOUT THE WORLD TO MAKE SUSTAINED OBSERVATIONS...

...IN THE REGIONS BEYOND THE EARTH'S ATMOSPHERE.

WHAT WAS NEVER ANNOUNCED PUBLICLY WAS EISENHOWER'S DECISION, EARLIER THAT YEAR,

TO AUTHORIZE SOMETHING CALLED WEAPONS SYSTEM-117L.

A COVERT PLAN TO LAUNCH THE WORLD'S FIRST SPY SATELLITE.

IN LIGHT OF THE AMERICAN ANNOUNCEMENT, THE SPACESHIPS ONCE DREAMED UP BY KOROLEV AND TIKHONRAVOV SUDDENLY SEEMED LESS LIKE FLIGHTS OF FANCY

AND MORE LIKE MILESTONES IN A NEW KIND OF COLD-WAR CONTEST:

A RACE TO SPACE.

WITHOUT A WORKING ROCKET, THAT FRONTIER REMAINED AS FAR-OFF AS EVER.

NEARLY A DECADE OF EXPERIMENTATION AND FALSE STARTS CULMINATED ON AUGUST 21, 1957:

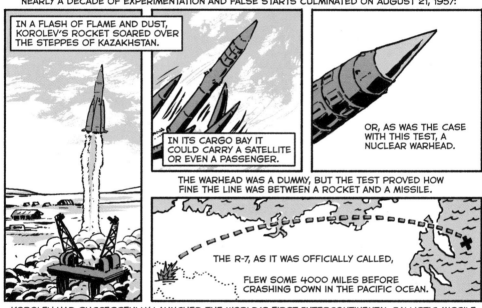

IN A FLASH OF FLAME AND DUST, KOROLEV'S ROCKET SOARED OVER THE STEPPES OF KAZAKHSTAN.

IN ITS CARGO BAY IT COULD CARRY A SATELLITE OR EVEN A PASSENGER.

OR, AS WAS THE CASE WITH THIS TEST, A NUCLEAR WARHEAD.

THE WARHEAD WAS A DUMMY, BUT THE TEST PROVED HOW FINE THE LINE WAS BETWEEN A ROCKET AND A MISSILE.

THE R-7, AS IT WAS OFFICIALLY CALLED,

FLEW SOME 4000 MILES BEFORE CRASHING DOWN IN THE PACIFIC OCEAN.

KOROLEV HAD SUCCESSFULLY LAUNCHED THE WORLD'S FIRST INTERCONTINENTAL BALLISTIC MISSILE.

TASS, THE STATE-RUN NEWS AGENCY, EXPLAINED THE TEST'S SIGNIFICANCE:

"THE RESULTS OBTAINED SHOW THAT THERE IS THE POSSIBILITY OF LAUNCHING MISSILES INTO ANY REGION OF THE TERRESTRIAL GLOBE."

FOR THE WEST, THIS WAS A TERRIFYING NEW DEVELOPMENT. ONE THAT WAS LARGELY IGNORED, SIMPLY BECAUSE IT SEEMED AS IF THE SOVIETS HAD TO BE EXAGGERATING.

MORE OF A PSYCHOLOGICAL STROKE THAN PROOF THAT RUSSIA HAS LEAPED AHEAD OF THE U.S. IN MISSILE DEVELOPMENT.

WAS THIS ANNOUNCEMENT SIMPLY DESIGNED TO CREATE FEAR AND HYSTERIA?

THE WORDS "SUCCESSFUL TEST" CAN MEAN MANY THINGS.

NO ONE WAS PREPARED FOR WHAT WOULD COME NEXT.

TIKHONRAVOV HAD BEEN WORKING ON A DREAM HE'D HATCHED MANY YEARS BEFORE:

THE "SIMPLEST SATELLITE."

TRUE TO ITS NAME, IT WAS A SMALL ALUMINUM ALLOY ORB, ABOUT 180 POUNDS, MOST OF WHICH WAS BATTERIES.

IT CARRIED TWO RADIO TRANSMITTERS TO BROADCAST INFORMATION ABOUT THE SATELLITE'S INTERNAL PRESSURE AND TEMPERATURE BACK TO SOVIET GROUND CONTROL.

ON OCTOBER 4, 1957, THE SATELLITE WAS LOADED INTO THE NOSE CONE OF KOROLEV'S ROCKET.

THE CHIEF DESIGNER HAD WAITED HIS WHOLE LIFE FOR THIS MOMENT.

3...2...

...1...

LIFTOFF!

I'M SEEING ELEVATED KEROSENE CONSUMPTION.

CORE STAGE SEPARATION.

I'M GETTING A CONFIRMATION SIGNAL FROM KAMCHATKA STATION.

HOLD OFF ON CELEBRATION. THE STATION PEOPLE COULD BE MISTAKEN.

LET'S JUDGE THE SIGNALS FOR OURSELVES WHEN THE SATELLITE COMES BACK AFTER ITS FIRST ORBIT AROUND THE EARTH.

FOR AN HOUR AND A HALF, KOROLEV AND HIS MEN SAT WAITING IN ANTICIPATION AS THEIR SATELLITE CIRCLED THE FAR SIDE OF THE GLOBE,

WAITING FOR A TELLTALE SOUND.

HELLO, MR. PRESIDENT, SORRY TO INTERRUPT YOUR VACATION.

BUT I THOUGHT YOU'D WANT TO HEAR A REPORT THAT JUST CAME OVER THE WIRES FROM MOSCOW...

ALL OVER THE WORLD, RADIOS AND TELEVISION SETS BROADCAST THE ALIEN, INCESSANT BEEPING OF SPUTNIK'S TRANSMITTERS. UNDENIABLE PROOF THAT THE SOVIETS HAD PLACED A SATELLITE IN ORBIT.

BEEEP...BEEEP...BEEEP...BEEEP...BE

LISTEN NOW FOR THE SOUND WHICH FOREVERMORE SEPARATES THE OLD FROM THE NEW.

WHILE THE SOVIET NEWS MEDIA WAS TRIUMPHANT...

"ARTIFICIAL EARTH SATELLITES WILL PAVE THE WAY TO INTERPLANETARY TRAVEL...WITNESS HOW THE FREED AND CONSCIENTIOUS LABOR OF THE PEOPLE OF THE NEW SOCIALIST SOCIETY MAKES THE MOST DARING DREAMS OF MANKIND A REALITY."

...THE EISENHOWER ADMINISTRATION TRIED TO DOWNPLAY THE NEWS WITH AN ANNOUNCEMENT BY HIS PRESS SECRETARY, JAMES HAGERTY.

OF COURSE IT IS OF GREAT SCIENTIFIC INTEREST, BUT IT DOES NOT COME AS ANY SURPRISE.

WE HAVE NEVER THOUGHT OF OUR PROGRAM AS IN A RACE WITH THE SOVIETS.

THE PRESS THOUGHT OTHERWISE.

RUSSIANS WIN RACE TO LAUNCH EARTH SATELLITE

SCIENTISTS MAP RED MOON'S ORBIT

THE SPACE AGE IS HERE

SOVIET MOON CIRCLING GLOBE

A PROPAGANDA TRIUMPH

THE AMERICAN REACTION TO SPUTNIK IS OFTEN PORTRAYED AS AN EXISTENTIAL CRISIS.

BUT ACCORDING TO POLLS FROM THE TIME, THE ATTENTION OF AMERICANS IN OCTOBER 1957 WAS MORE FRACTURED.

TOP OF THE NINTH AT YANKEE STADIUM IN GAME SEVEN OF THE WORLD SERIES, BASES ARE LOADED...

MEMBERS OF THE 101ST AIRBORNE PROTECT NEGRO STUDENTS ON THEIR WAY TO SCHOOL...

STILL, SPUTNIK MADE FOR GOOD COPY, EVEN WEEKS AFTER THE LAUNCH.

Let us not pretend that Sputnik is anything but a defeat for the United States...

LIFE

SAVVY POLITICIANS, LIKE SENATOR LYNDON B. JOHNSON, STOKED THE FLAMES OF PARANOIA:

SOON THEY WILL BE DROPPING BOMBS ON US FROM SPACE LIKE KIDS DROPPING ROCKS ONTO CARS FROM FREEWAY OVERPASSES.

IT DIDN'T HELP THAT EISENHOWER HIMSELF EXHIBITED COMPLETE INDIFFERENCE TO THE SOVIET SATELLITE.

NOW, SO FAR AS THE SATELLITE ITSELF IS CONCERNED, THAT DOES NOT RAISE MY APPREHENSIONS, NOT ONE IOTA.

THE MERE FACT THAT THIS THING ORBITS INVOLVES NO NEW DISCOVERY TO SCIENCE...

...SO IN ITSELF IT IMPOSES NO ADDITIONAL THREAT TO THE UNITED STATES.

IN TRUTH, EISENHOWER WASN'T ALARMED BY SPUTNIK, BECAUSE SPUTNIK WASN'T A SURPRISE.

THE CIA HAD WARNED HIM THAT THE SOVIETS WERE ABOUT TO LAUNCH A SATELLITE.

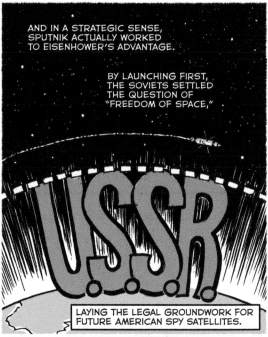

AND IN A STRATEGIC SENSE, SPUTNIK ACTUALLY WORKED TO EISENHOWER'S ADVANTAGE.

BY LAUNCHING FIRST, THE SOVIETS SETTLED THE QUESTION OF "FREEDOM OF SPACE,"

LAYING THE LEGAL GROUNDWORK FOR FUTURE AMERICAN SPY SATELLITES.

THE PRESIDENT'S ADVISERS OFFERED OTHER REASONS NOT TO WORRY.

PUTTING AN OBJECT IN ORBIT IS FAR LESS COMPLICATED THAN AIMING A MISSILE AT A TARGET HALFWAY ACROSS THE GLOBE.

IT'S JUST NOT REALISTIC TO DEPLOY NUCLEAR WEAPONS FROM A SATELLITE.

BUT WHAT DID CATCH EISENHOWER OFF GUARD WAS HOW IMMEDIATELY AND INDELIBLY SPUTNIK BECAME A SYMBOL

OF SOVIET SUPERIORITY.

OF AMERICAN COMPLACENCE.

DECEMBER 6, 1957

CAPE CANAVERAL, FLORIDA

TELEVISION CREWS PREPARED TO BROADCAST THE MOMENTOUS LAUNCH OF THE FIRST AMERICAN SATELLITE.

3...2....

...1...

LIFTOFF.

PROJECT VANGUARD WAS SUPPOSED TO HAVE LAUNCHED MONTHS EARLIER, BEATING SPUTNIK TO ORBIT.

BUT THERE HAD BEEN A STRING OF DESIGN CHANGES AND DELAYS.

EISENHOWER HOPED THE LAUNCH WOULD SOOTHE THE ANGST CAUSED BY SPUTNIK.

HE WAS WRONG.

JUST AS THE SOVIET PRESS HAD GIVEN SPUTNIK ITS NAME,

AMERICAN NEWSPAPERS CHRISTENED THEIR HOMEGROWN SATELLITE ON ITS MAIDEN VOYAGE INTO THE SOUTH FLORIDA SCRUB:

"FLOPNIK"

VON BRAUN WAS FURIOUS.

HE'D BEEN SIDELINED FROM THE SATELLITE PROJECT BECAUSE HE WORKED FOR THE ARMY.

WE CAN PUT UP A SATELLITE IN 60 DAYS. JUST GIVE US THE GREEN LIGHT AND 60 DAYS!

TRUE TO HIS WORD, ON JANUARY 31, 1958, VON BRAUN'S ROCKET LIFTED THE FIRST U.S. SATELLITE INTO ORBIT.

IF THERE WAS AN INAUGURAL YEAR TO THE SPACE AGE, IT WAS 1958.

THE SOVIETS AND THE AMERICANS ATTEMPTED 22 LAUNCHES THAT YEAR.

EIGHT OF WHICH WERE SUCCESSFUL.

IN JULY 1958, EISENHOWER SIGNED INTO LAW THE NATIONAL AERONAUTICS AND SPACE ACT.

FROM HERE ON, SPACE WOULD BE THE PURVIEW OF A SINGLE GOVERNMENT AGENCY, A CIVILIAN ORGANIZATION:

NASA.

MEANWHILE, THE SOVIET LEADERSHIP WANTED MORE SPUTNIKS,

MORE OPPORTUNITIES TO COMMAND THE FEAR AND RESPECT OF THE WEST.

BUT IN ALL THAT PUBLICITY, THERE WERE THREE WORDS THAT NEVER APPEARED IN PRINT: SERGEI PAVLOVICH KOROLEV.

THE MAN WHO MADE SPUTNIK POSSIBLE--

WHO OVER THE NEXT FEW YEARS WOULD ENGINEER A STREAK OF SOVIET FIRSTS IN THE SPACE RACE--

BEGAN TO DISAPPEAR.

HIS NAME WAS QUIETLY ERASED FROM THE OFFICIAL RECORD.

AND IN HIS PLACE, A PHYSICIST NAMED LEONID SEDOV BECAME THE PUBLIC FACE OF THE SOVIET SPACE PROGRAM.

NO MATTER. SPUTNIK HAD ALWAYS BEEN FOR KOROLEV A WAYPOINT ON A LONGER, MORE AMBITIOUS JOURNEY.

I CALL IT MEЧTA.

"THE DREAM"

THOUGH THE OFFICIAL NAME FOR KOROLEV'S MOON PROBE WAS "LUNA."

LUNA 1 WAS MEANT TO CRASH INTO THE MOON, BUT MISSED.

LUNA 2 HIT ITS TARGET, RELEASING A BARRAGE OF STAINLESS STEEL BADGES ENGRAVED WITH THE SOVIET COAT OF ARMS.

CCCP

AND LUNA 3...

OCTOBER 18, 1959

RECEIVING A STRONGER SIGNAL NOW.

IT LOOKS LIKE WE'RE GETTING MORE THAN JUST STATIC.

WHILE THE AMERICAN MILITARY SECRETLY DEVISED AN EARTH-CIRCLING SATELLITE

TO PHOTOGRAPH THE HIDDEN REACHES OF THE SOVIET UNION,

KOROLEV'S "DREAM" PROBE WAS SURVEILLING A LAND NO HUMAN HAD EVER SEEN.

THE CHIEF DESIGNER HAD STOLEN A GLIMPSE OF THE FAR SIDE OF THE MOON.

119

CHAPTER XI

COLUMBIA

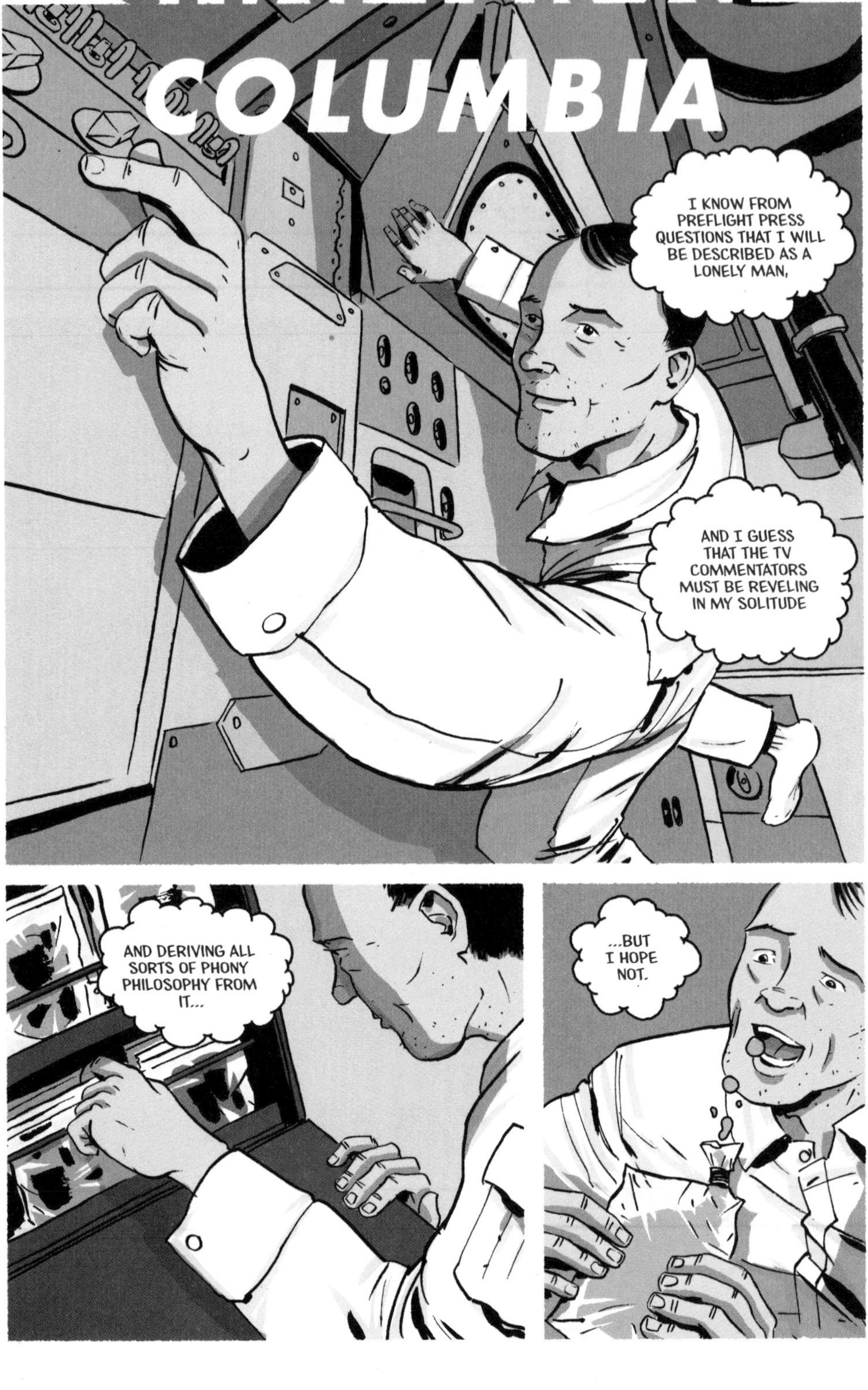
I KNOW FROM PREFLIGHT PRESS QUESTIONS THAT I WILL BE DESCRIBED AS A LONELY MAN,
AND I GUESS THAT THE TV COMMENTATORS MUST BE REVELING IN MY SOLITUDE
AND DERIVING ALL SORTS OF PHONY PHILOSOPHY FROM IT...
...BUT I HOPE NOT.

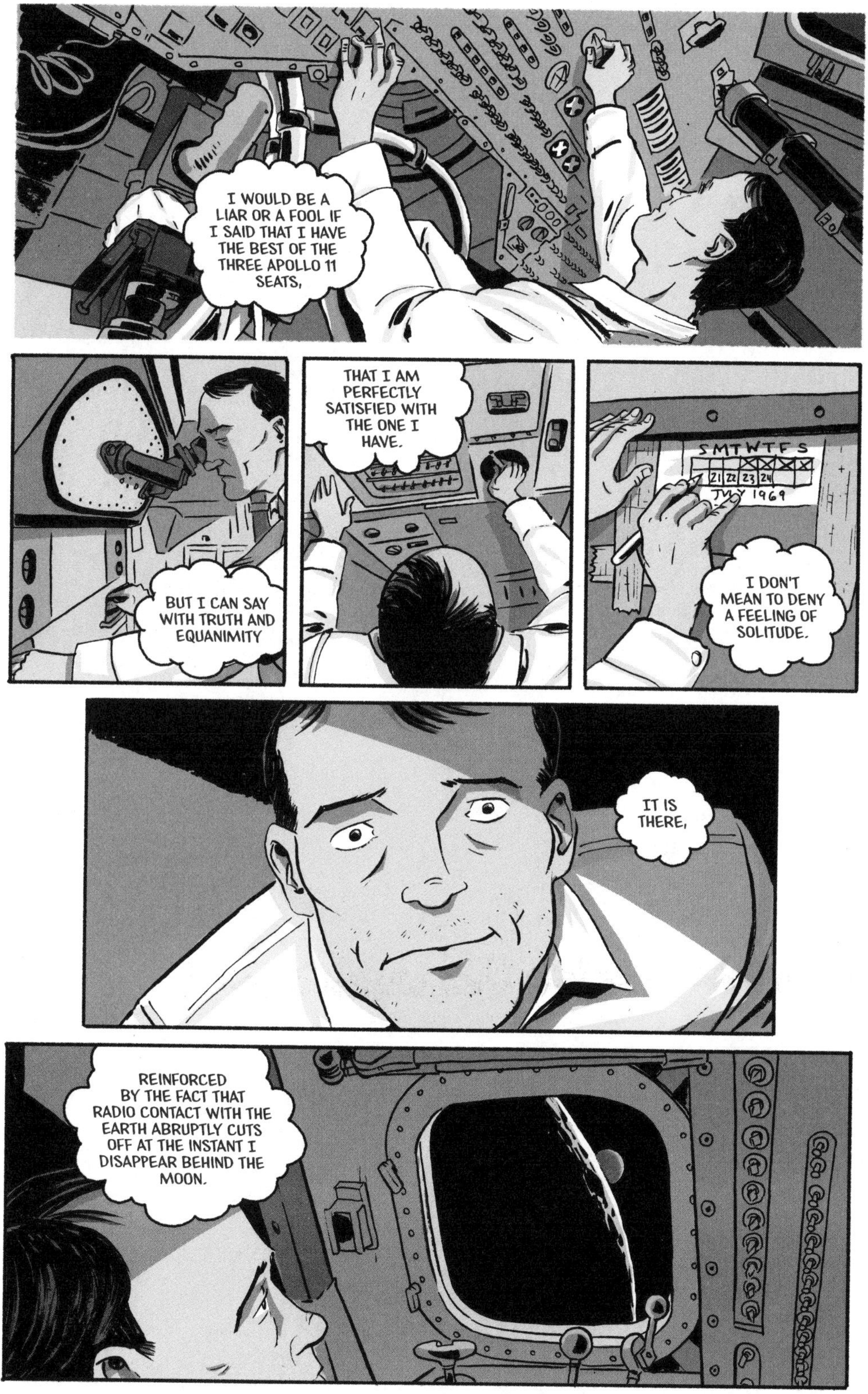
I WOULD BE A LIAR OR A FOOL IF I SAID THAT I HAVE THE BEST OF THE THREE APOLLO 11 SEATS,
BUT I CAN SAY WITH TRUTH AND EQUANIMITY
THAT I AM PERFECTLY SATISFIED WITH THE ONE I HAVE.
S M T W T F S
21 22 23 24
JULY 1969
I DON'T MEAN TO DENY A FEELING OF SOLITUDE.
IT IS THERE,
REINFORCED BY THE FACT THAT RADIO CONTACT WITH THE EARTH ABRUPTLY CUTS OFF AT THE INSTANT I DISAPPEAR BEHIND THE MOON.

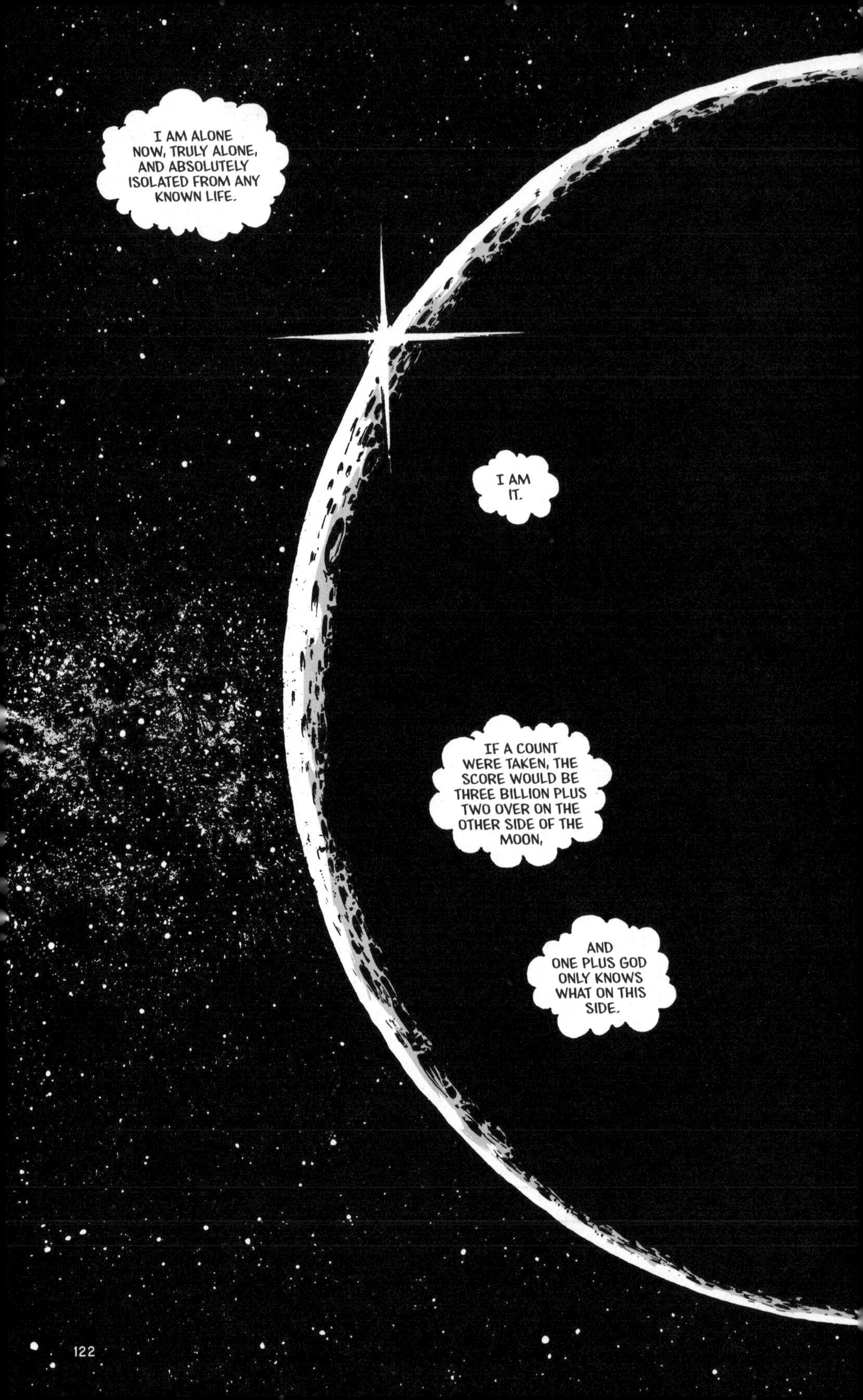
I AM ALONE NOW, TRULY ALONE, AND ABSOLUTELY ISOLATED FROM ANY KNOWN LIFE.
I AM IT.
IF A COUNT WERE TAKEN, THE SCORE WOULD BE THREE BILLION PLUS TWO OVER ON THE OTHER SIDE OF THE MOON,
AND ONE PLUS GOD ONLY KNOWS WHAT ON THIS SIDE.

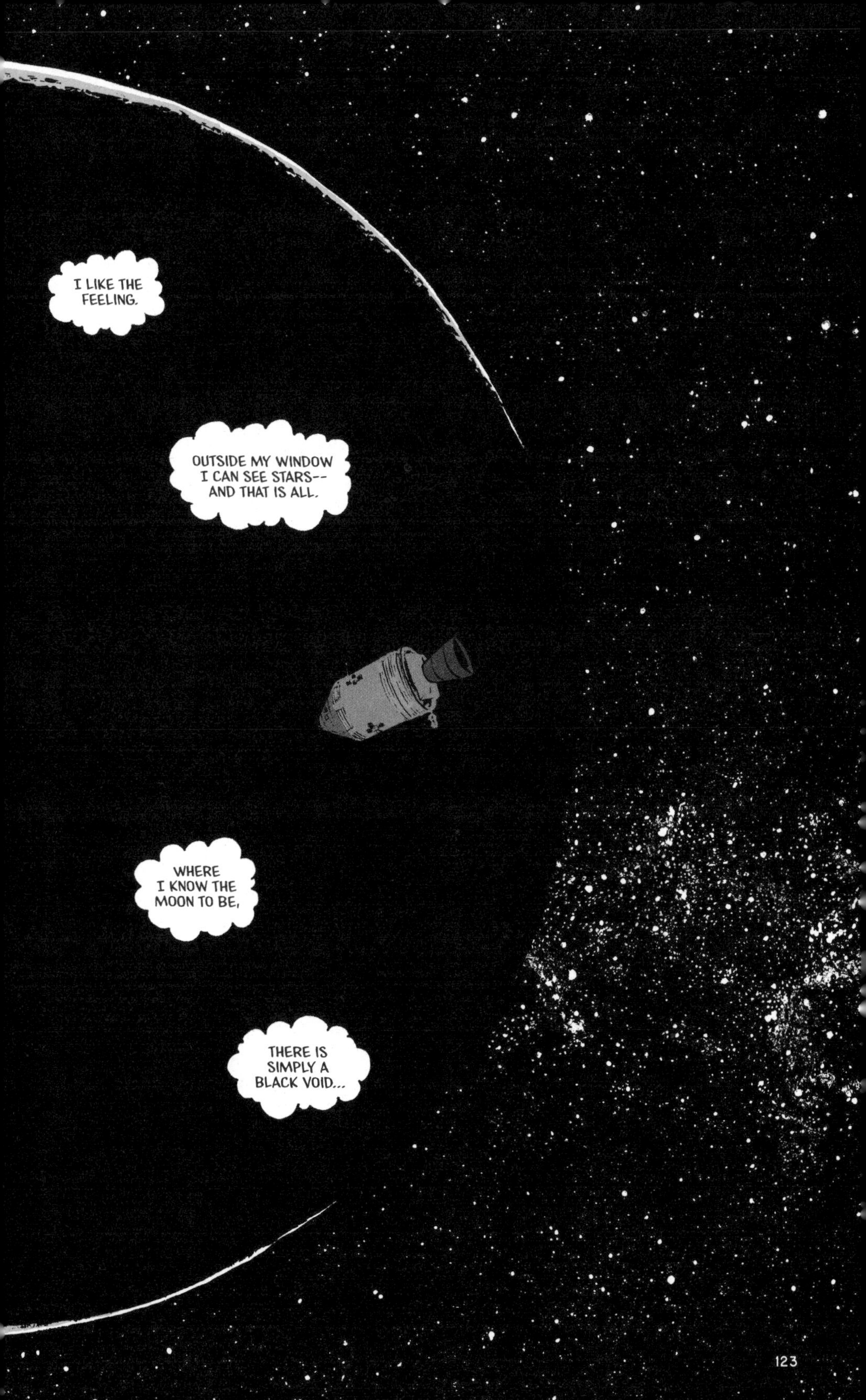
I LIKE THE FEELING.
OUTSIDE MY WINDOW I CAN SEE STARS--AND THAT IS ALL.
WHERE I KNOW THE MOON TO BE,
THERE IS SIMPLY A BLACK VOID...

CHAPTER XII
ASTRONAUTS

A MAN SHOULD HAVE THE ABILITY TO GO UP IN A HURTLING
PIECE OF MACHINERY AND PUT HIS HIDE ON THE LINE AND
THEN HAVE THE MOXIE, THE REFLEXES, THE EXPERIENCE,
THE COOLNESS, TO PULL IT BACK IN THE LAST YAWNING
MOMENT--AND THEN TO GO UP AGAIN *THE NEXT DAY*, AND
THE NEXT DAY, AND EVERY NEXT DAY, EVEN IF THE SERIES
SHOULD PROVE INFINITE--AND, ULTIMATELY, IN ITS BEST
EXPRESSION, DO SO IN A CAUSE THAT MEANS SOMETHING
TO THOUSANDS, TO A PEOPLE, A NATION, TO HUMANITY,
TO GOD.

--TOM WOLFE

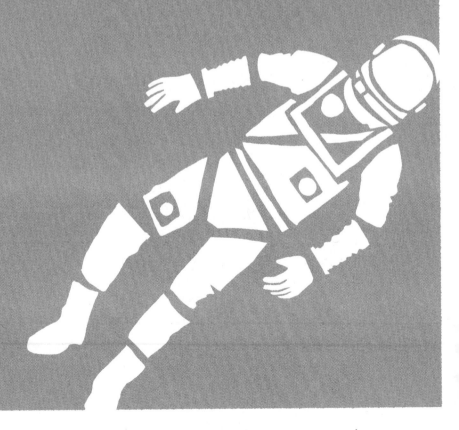

SIXTY-THREE COMBAT MISSIONS IN WORLD WAR II.

SIX THOUSAND HOURS IN A COCKPIT.

SAY "AHH."

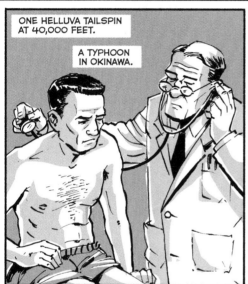

ONE HELLUVA TAILSPIN AT 40,000 FEET.

A TYPHOON IN OKINAWA.

A NOSE BROKEN SO MANY TIMES IT'S HARD TO FIT INTO AN OXYGEN MASK.

WHAT'S THAT FOR?

WE NEED TO SAMPLE YOUR STOMACH ACIDS.

ALL OF IT WAS NOTHING NEXT TO THIS.

JUST IMAGINE YOU'RE SWALLOWING A LONG PIECE OF SPAGHETTI.

ON HIS GRAVESTONE THEY'LL WRITE: "DONALD K. SLAYTON: CHOKED ON A RUBBER HOSE"

GOOD AFTERNOON, MR. SLAYTON.

HOPE THAT WASN'T TOO BAD.

...COUGH... AFTERNOON, DOC.

DR. RANDY LOVELACE HAD BEEN AT THIS LONG ENOUGH TO KNOW THAT HIS PATIENTS NEVER COMPLAINED.

AT LEAST NOT TO HIS FACE.

THERE WEREN'T MANY TESTS THAT DR. LOVELACE--HEAD OF LIFE SCIENCES AT NASA-- HADN'T AT SOME POINT TRIED ON HIMSELF.

IN 1939, TO PROVE HIS NEW DESIGN FOR AN OXYGEN MASK, HE JUMPED OUT OF AN AIRPLANE, BRIEFLY PASSED OUT, AND WOULD HAVE FALLEN TO HIS DEATH IF NOT FOR HIS MASK.

IT WAS HIS FIRST TIME IN A PARACHUTE. HE SET AN ALTITUDE RECORD THAT DAY. AND HIS INVENTION WORKED.

DR. LOVELACE PIONEERED THE STUDY OF WHAT HAPPENS TO THE HUMAN BODY AT ALTITUDE.

BY 1958, AT THE HEIGHT OF THE SPACE RACE, HE WAS IN HIGH DEMAND.

LOVELACE CLINIC

SO THE MAIN REASON DONALD SLAYTON--DEKE TO HIS FELLOW AIR FORCE TEST PILOTS--WOULD NEVER COMPLAIN TO DR. LOVELACE...

WHAT'S NEXT, DOC?

...WAS BECAUSE DEKE WANTED TO GO TO SPACE.

NASA HEADQUARTERS
WASHINGTON, D.C.

THE NAME'S SLAYTON, DONALD SLAYTON.

GO ON IN. THE BRIEFING WILL BEGIN SHORTLY.

IN 1959 NASA SENT OUT A CALL TO MILITARY TEST PILOTS FOR PROJECT ASTRONAUT.

ALTHOUGH THAT NAME WAS LATER DROPPED.

WILL SOMEONE GET THE LIGHTS?

WE'RE PREPARING A SERIES OF SPACEFLIGHTS, BOTH MANNED AND UNMANNED, UNDER THE NAME PROJECT MERCURY.

A SMALL CAPSULE, LARGE ENOUGH FOR ONE ASTRONAUT, WILL MAKE SUBORBITAL FLIGHTS USING THE ARMY'S REDSTONE ROCKET...

...AND, WITH THE USE OF THE AIR FORCE'S LARGER ATLAS ROCKET, WILL EVENTUALLY BE PUT INTO ORBIT.

SO IF THEY CAN FLY THIS THING WITHOUT PILOTS,

WHY DO THEY EVEN NEED US IN THE FIRST PLACE?

OUR PRIMARY OBJECTIVE WITH THIS SERIES OF LAUNCHES...

...IS TO TEST HOW THE ENVIRONMENT OF SPACE AFFECTS THE HUMAN BODY.

WHICH BRINGS US BACK TO THE RUBBER HOSE.

ONE OF THE MANY INDIGNITIES TO WHICH PROJECT MERCURY HOPEFULS WERE SUBJECTED.

THIS IS GOING TO FEEL A LITTLE STRANGE.

SPLURT

DEKE SLAYTON AND 31 OTHER TEST PILOTS

TRAVELED TO DR. LOVELACE'S CLINIC IN ALBUQUERQUE, NEW MEXICO, AND THEN TO A SPECIAL LAB IN DAYTON, OHIO,

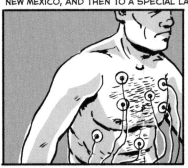

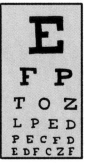

E
F P
T O Z
L P E D
P E C F D
E D F C Z F

TO PARTICIPATE IN A BATTERY OF TESTS TO DETERMINE EACH MAN'S OVERALL FITNESS FOR THE JOB,

AND TO PUSH THE VERY LIMITS OF WHAT A HUMAN MIGHT BE EXPECTED TO ENDURE.

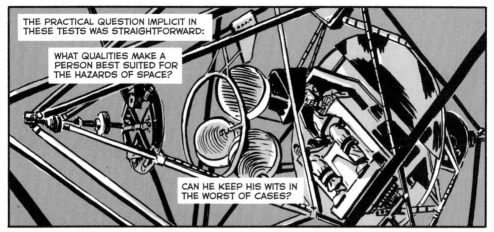

THE PRACTICAL QUESTION IMPLICIT IN THESE TESTS WAS STRAIGHTFORWARD:

WHAT QUALITIES MAKE A PERSON BEST SUITED FOR THE HAZARDS OF SPACE?

CAN HE KEEP HIS WITS IN THE WORST OF CASES?

CAN HE HANDLE SEVERE HEAT AND COLD?

CAN HE ENDURE THE EXTREME ACCELERATION OF A ROCKET LAUNCH?

PSYCHIATRISTS PROBED THE MEN FOR ANY SIGN OF MENTAL FRAILTY.

WHO ARE YOU?

COMPLETE THIS SENTENCE: "MY MOTHER IS..."

TRUE OR FALSE: STRANGERS KEEP TRYING TO HURT ME.

WHO ARE YOU?

BUT THERE WAS NO EASY ANSWER TO THE QUESTION AT THE HEART OF PROJECT MERCURY: WHAT MAKES AN ASTRONAUT?

IT'S A MADE-UP WORD, FROM GREEK, MEANING "STAR-SAILOR."

AN ECHO OF THE MYTHIC ARGONAUTS, AND THEIR QUEST FOR THE GOLDEN FLEECE.

LIKE THOSE LEGENDARY ADVENTURERS,

THE MODERN ASTRONAUTS WOULD NAVIGATE UNCHARTED EXPANSES RIFE WITH DANGER AND GLORY.

FOR THAT, THEY WOULD NEED SOMETHING MORE THAN SKILL AND STAMINA.

ALTHOUGH NO ONE QUITE REALIZED IT YET, NASA WAS LOOKING FOR *HEROES*.

AND HEROES ARE EXACTLY WHAT THE AMERICAN PEOPLE SAW ON APRIL 9, 1959,

WHEN SEVEN MEN SAUNTERED INTO A NASA PRESS CONFERENCE.

LADIES AND GENTLEMEN...THE NATION'S PROJECT MERCURY ASTRONAUTS.

WHAT DO YOUR WIVES THINK ABOUT ALL THIS?

WILL THE RUSSIANS BEAT US TO SPACE?

WHY FAMILY MEN INSTEAD OF BACHELORS?

WHAT ARE YOUR IQs?

DO YOU HAVE A SUSTAINING RELIGIOUS FAITH?

WHICH OF THE TESTS WAS HARDEST?

WHAT MOTIVATED YOU TO BECOME AN ASTRONAUT?

THEY DID THEIR BEST TO ANSWER THE QUESTIONS.

BUT THESE WERE MEN MORE AT EASE IN A FIGHTER JET THAN IN A ROOM FULL OF REPORTERS.

THIS IS AN EXCELLENT OPPORTUNITY TO BE IN ON SOMETHING NEW.

I THINK IN MY ANSWER TO WHAT MY MOTIVATION IS, I THINK IT IS TYPICAL OF MOST OF US IN THIS COUNTRY: WE ARE INTERESTED IN NEW THINGS.

I AM JUST GRATEFUL FOR AN OPPORTUNITY TO SERVE IN THIS CAPACITY.

JOHN GLENN WAS DIFFERENT.

HE WAS BORN FOR THE SPOTLIGHT.

EVERY ONE OF US WOULD FEEL GUILTY I THINK IF WE DIDN'T MAKE THE FULLEST USE OF OUR TALENTS IN VOLUNTEERING FOR SOMETHING THAT IS AS IMPORTANT AS THIS IS TO OUR COUNTRY AND THE WORLD IN GENERAL RIGHT NOW.

BEING AN ASTRONAUT MEANT PATRIOTISM AND SACRIFICE.

IT WASN'T A JOB.

IT WAS A HIGHER CALLING.

THE PRESS CONFERENCE MARKED A MOMENT OF TRANSFORMATION FOR THESE SEVEN MEN:

FROM TEST PILOTS--ELITE, BUT MOSTLY ANONYMOUS OUTSIDE OF THE MILITARY--TO NATIONAL CELEBRITIES.

THEY'RE APPLAUDING US LIKE WE'VE ALREADY DONE SOMETHING, LIKE WE'RE HEROES.

BY THE END OF THE DAY, AL, GUS, GORDO, JOHN, WALLY, SCOTT, AND DEKE WERE HOUSEHOLD NAMES.

THE NEW JOB, HOWEVER, DIDN'T PAY ANY BETTER THAN THE OLD ONE.

LIFE MAGAZINE OFFERED SOME RELIEF, IN THE FORM OF A HEFTY CONTRACT IN EXCHANGE FOR THE EXCLUSIVE RIGHT TO THEIR STORIES.

LIFE REPORTERS SHADOWED THE "MERCURY 7" AT HOME AND AT WORK.

WHAT CAME OUT IN THE GLOSSY PHOTO SPREADS WAS A VISION OF THE 1950s AMERICAN DREAM:

THE DEMURE, SUPPORTIVE WIFE

THE BRIGHT, WELL-MANNERED CHILD

AND AT THE HEAD OF THE HOUSEHOLD, THE VERY DEFINITION OF MASCULINITY:

THE ASTRONAUT EVERYMAN.

BUT WAS THE WORK OF AN ASTRONAUT REALLY SOMETHING ONLY A MAN COULD DO?

LOOK

SHOULD A GIRL BE FIRST IN SPACE?

ONE OF THE FEW OUTSPOKEN ADVOCATES FOR SENDING WOMEN INTO SPACE WAS DR. RANDY LOVELACE.

CERTAIN QUALITIES OF THE FEMALE SPACE PILOT ARE PREFERABLE TO THOSE OF HER MALE COLLEAGUE.

ON AVERAGE, WOMEN WERE SMALLER THAN MEN AND CONSUMED FEWER CALORIES AND LESS OXYGEN, WHICH WAS IDEAL WHEN THE GOAL WAS TO BUILD AS LIGHT A SPACECRAFT AS POSSIBLE.

IN LATE 1959, LOVELACE TEAMED UP WITH BRIGADIER GENERAL DONALD FLICKINGER, WHO HAD BEEN PLANNING HIS OWN "GIRL ASTRONAUT" PROGRAM FOR THE AIR FORCE.

I DON'T SEE WHY WE SHOULDN'T PUT A GROUP OF LADIES THROUGH THE SAME TESTS AS THE MERCURY BOYS.

IT WON'T BE AN OFFICIAL PART OF NASA, OF COURSE, BUT WE CAN USE MY CLINIC.

THEY STUMBLED UPON THEIR FIRST TEST SUBJECT AFTER A SWIM AT MIAMI BEACH,

DURING A LULL AT AN AVIATION CONFERENCE.

PLEASED TO MEET YOU. I'M JERRIE COBB.

THE FACT THAT WOMEN WEREN'T ALLOWED TO FLY IN THE MILITARY MAKES GERALDYN "JERRIE" COBB'S CAREER ALL THE MORE REMARKABLE.

SHE STARTED FLYING AT THE AGE OF 12, SET SEVERAL ALTITUDE RECORDS, AND ACCRUED MORE THAN 10,000 HOURS IN A COCKPIT.

LOVELACE HAD FOUND HIS IDEAL CANDIDATE.

IN FEBRUARY 1960--ALMOST A YEAR AFTER HE HAD FIRST TESTED THE MERCURY VOLUNTEERS--THE 29-YEAR-OLD JERRIE COBB ARRIVED IN ALBUQUERQUE.

IN TEST AFTER TEST, COBB MET OR EXCEEDED THE BENCHMARKS SET BY THE MEN.

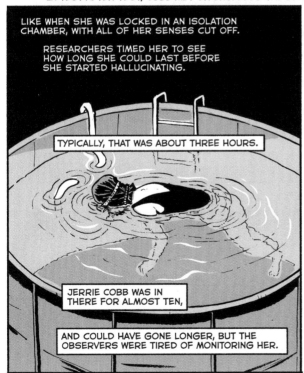

LIKE WHEN SHE WAS LOCKED IN AN ISOLATION CHAMBER, WITH ALL OF HER SENSES CUT OFF.

RESEARCHERS TIMED HER TO SEE HOW LONG SHE COULD LAST BEFORE SHE STARTED HALLUCINATING.

TYPICALLY, THAT WAS ABOUT THREE HOURS.

JERRIE COBB WAS IN THERE FOR ALMOST TEN,

AND COULD HAVE GONE LONGER, BUT THE OBSERVERS WERE TIRED OF MONITORING HER.

DR. LOVELACE ANNOUNCED THE RESULTS OF HIS TESTING AT A CONFERENCE IN SWEDEN.

MISS COBB IS QUALIFIED TO LIVE, OBSERVE, AND DO OPTIMAL WORK IN THE ENVIRONMENT OF SPACE, AND RETURN SAFELY TO EARTH.

WHEN LOVELACE ANNOUNCED HIS RESULTS, NEWSPAPERS ACROSS THE WORLD TRUMPETED THE STORY OF THE FIRST "ASTRONAUTRIX."

(OR "FEMINAUT" OR "ASTRO-NETTE." NO ONE COULD AGREE ON WHAT TO CALL HER.)

DESPITE HER QUALIFICATIONS, COBB WAS DESCRIBED IN TERMS USUALLY RESERVED FOR THE STYLE SECTION.

"MEASUREMENTS: 36-27-34."

"A SLENDER BLONDE."

"SAID SHE LOST SEVEN POUNDS DURING A WEEK OF TESTING."

"SHE'S THE BEST-ORGANIZED GIRL PILOT I'VE EVER SEEN-- AND THE MOST FEMININE."

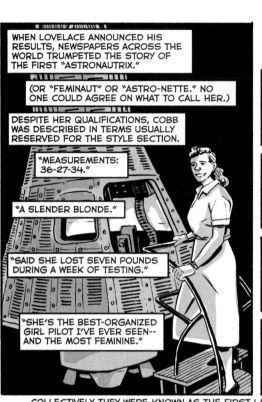

INSPIRED BY THE NEWS COVERAGE, DOZENS OF WOMEN VOLUNTEERED FOR LOVELACE'S "WOMAN IN SPACE" PROGRAM. ALL TOLD, OF THE 25 CANDIDATES INVITED, 13 PASSED.

COLLECTIVELY THEY WERE KNOWN AS THE FIRST LADY ASTRONAUT TRAINEES, OR F.L.A.T.S.

NASA IGNORED THEM, BUT THE PRESS LOVED THEIR STORY. AMID ALL THE PUBLICITY AND PROFILES, COBB STARTED TO BELIEVE THAT SHE MIGHT REALLY ONE DAY GO INTO ORBIT.

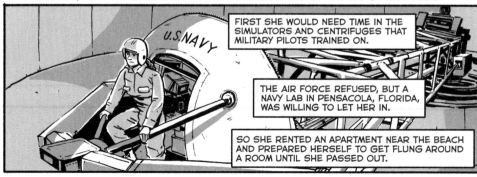

FIRST SHE WOULD NEED TIME IN THE SIMULATORS AND CENTRIFUGES THAT MILITARY PILOTS TRAINED ON.

THE AIR FORCE REFUSED, BUT A NAVY LAB IN PENSACOLA, FLORIDA, WAS WILLING TO LET HER IN.

SO SHE RENTED AN APARTMENT NEAR THE BEACH AND PREPARED HERSELF TO GET FLUNG AROUND A ROOM UNTIL SHE PASSED OUT.

HER SUCCESS ON THE SIMULATORS ENCOURAGED THE OTHER F.L.A.T.S TO PLAN TRIPS TO PENSACOLA.

THEN, ONE MORNING IN 1961, COBB GOT A TELEGRAM FROM LOVELACE.

"REGRET TO ADVISE...PENSACOLA CANCELED...WILL NOT BE POSSIBLE TO CARRY OUT THIS PART OF PROGRAM..."

THERE WOULD BE NO WOMEN IN SPACE.

AT LEAST NOT IN AMERICA.

THE REASONS OFFERED WERE VAGUE. "LACK OF OFFICIAL REQUEST."

LET'S STOP THIS.

NOW!

NOBODY WITH ANY CLOUT THOUGHT FEMALE ASTRONAUTS WERE A GOOD IDEA.

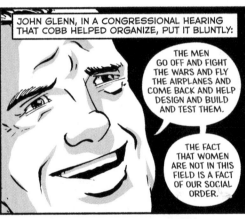

JOHN GLENN, IN A CONGRESSIONAL HEARING THAT COBB HELPED ORGANIZE, PUT IT BLUNTLY:

THE MEN GO OFF AND FIGHT THE WARS AND FLY THE AIRPLANES AND COME BACK AND HELP DESIGN AND BUILD AND TEST THEM.

THE FACT THAT WOMEN ARE NOT IN THIS FIELD IS A FACT OF OUR SOCIAL ORDER.

NASA NEVER SERIOUSLY CONSIDERED SENDING WOMEN TO SPACE. THE AGENCY ALREADY HAD SEVEN SUPERSTAR ASTRONAUTS.

DR. VON BRAUN, WILL THERE BE WOMEN IN SPACE?

ONLY IF NASA GIVES US AN EXTRA 110 POUNDS OF PAYLOAD FOR "RECREATIONAL EQUIPMENT."

THE UNITED STATES LOST YET ANOTHER LEG OF THE SPACE RACE TO THE SOVIETS WHEN, ON JUNE 16, 1963, THE COSMONAUT VALENTINA TERESHKOVA BECAME THE FIRST WOMAN IN ORBIT.

IT WOULD TAKE NASA 20 MORE YEARS TO CATCH UP.

135

THROUGHOUT JERRIE COBB'S ORDEAL, THE MERCURY ASTRONAUTS WERE MOSTLY KEEPING THEIR HEADS DOWN, DEVELOPING THE TECHNIQUES OF SPACEFLIGHT.

CLACK

RETROMANUAL FUSE SWITCH IS OFF...

AT SEA, THEY PRACTICED WHAT TO DO IF THE MISSION WENT AS PLANNED...

...AND IN THE DESERT, THEY LEARNED HOW TO SURVIVE IF THE MISSION WENT AWRY.

THE LARGER-THAN-LIFE PUBLIC IMAGE THAT FOLLOWED THESE MEN WHEREVER THEY WENT WAS HELPFUL FOR NASA IN MARSHALING SUPPORT FOR THE SPACE PROGRAM.

BUT THAT CELEBRITY WASN'T WHAT MADE THESE MEN ASTRONAUTS.

NOR WAS IT NECESSARILY THE GRIT THEY'D SHOWN DURING THE LOVELACE TESTS.

NO, THESE MEN WERE ASTRONAUTS BECAUSE OF WHAT THEY'D DONE IN THEIR OLD LIVES AS TEST PILOTS.

WHERE THEIR JOB WAS TO TRANSLATE THE VISCERAL EXPERIENCE OF PLANE FLIGHT INTO TERMS THAT ENGINEERS AND MECHANICS COULD WORK WITH.

ALL THE WHILE, NO ONE WAS SURE WHO WOULD GET TO GO UP FIRST.

I WANT YOU GUYS TO TAKE A VOTE...

THE FINAL DECISION WAS UP TO DR. ROBERT GILRUTH, DIRECTOR OF THE MANNED SPACECRAFT CENTER AT CAPE CANAVERAL, NASA'S LAUNCH SITE.

BUT HE WASN'T GIVING ANY CLUES.

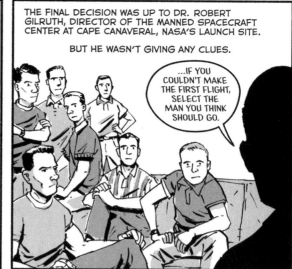

...IF YOU COULDN'T MAKE THE FIRST FLIGHT, SELECT THE MAN YOU THINK SHOULD GO.

AND THEN, ONE AFTERNOON IN JANUARY 1961, GILRUTH CALLED THE ASTRONAUTS INTO HIS OFFICE.

SHEPARD WILL MAKE THE FIRST SUBORBITAL FLIGHT, THEN GRISSOM.

GLENN, YOU HAVE THE FIRST ORBITAL FLIGHT.

NO ONE SAYS A WORD UNTIL THE OFFICIAL ANNOUNCEMENT.

THE LAUNCH DATE WAS SET FOR THE FIRST WEEK OF MAY, 1961.

A HISTORIC ACHIEVEMENT, THE CULMINATION OF YEARS OF EFFORT, MILLIONS OF DOLLARS AND--

POYEKHALI!

ON APRIL 12, THE SOVIET COSMONAUT YURI GAGARIN ROCKETED INTO ORBIT.

ONCE AGAIN, NASA WAS TOO LATE.

AS GAGARIN MADE HIS WAY AROUND THE EARTH, SERGEI KOROLEV, BACK AT THE LAUNCH SITE, KEPT TABS ON THE YOUNG COSMONAUT.

HOW DO YOU FEEL?

I SEE THE EARTH. I SEE THE CLOUDS...IT'S... *BEAUTIFUL.*

IT'S HARD TO SAY WHO WAS MORE SURPRISED THAT DAY:

THE PEASANTS WHO SAW A MYSTERIOUS VISITOR FALL FROM THE SKY...

...OR NASA'S SPOKESMAN JOHN "SHORTY" POWERS, WHO GOT A LATE-NIGHT PHONE CALL FROM A REPORTER.

"GOOD MORNING" MY ASS. WHATTYA WANT?

NASA'S REACTION TO THE RUSSIANS ORBITING A COSMONAUT?

#$&% YOU, BARBREE. WE'RE ASLEEP HERE!

JAY BARBREE, A CORRESPONDENT FOR NBC, IMPISHLY REPORTED POWERS'S RESPONSE:

OVERNIGHT, THE RUSSIANS PUT A MAN INTO SPACE, AND COLONEL JOHN POWERS, THE SPOKESMAN FOR THE MERCURY 7 ASTRONAUTS, TELLS ME "NASA'S ASLEEP."

THE SPACE AGENCY WILL WAIT TO HEAR ABOUT MAN'S FIRST FLIGHT INTO EARTH ORBIT OVER EGGS AND BACON.

138

FIVE DAYS AFTER GAGARIN'S SHOCKING ORBIT, PRESIDENT JOHN F. KENNEDY LEARNED THAT A CIA OPERATION HE'D SIGNED OFF ON--TO OVERTHROW FIDEL CASTRO IN CUBA--WAS A FIASCO.

THE COMMUNIST REVOLUTIONARY CHE GUEVARA SENT KENNEDY A NOTE IN THE AFTERMATH OF THE BAY OF PIGS DEBACLE.

THE PRESIDENT NEEDED SOME GOOD NEWS.

I SIMPLY THANKED HIM. BEFORE THE INVASION, THE REVOLUTION WAS WEAK. NOW IT'S STRONGER THAN EVER.

ALAN SHEPARD'S FLIGHT HELPED. HE MADE IT TO SPACE AND BACK WITHOUT A HITCH.

AS DID THE NEXT MERCURY FLIGHT, PILOTED BY VIRGIL "GUS" GRISSOM,

THOUGH A FOUL-UP WITH THE HATCH AFTER REENTRY SENT THE CAPSULE TO THE BOTTOM OF THE ATLANTIC.

AND THEN, IN EARLY 1962, THE BIG ONE: JOHN GLENN'S ORBITAL FLIGHT.

ZERO G AND I FEEL FINE.

A MALFUNCTIONING SENSOR CUT HIS TRIP SHORT,

BUT HE STILL MANAGED THREE CIRCUITS AROUND THE EARTH.

HE RETURNED TO A HERO'S WELCOME.

NEVER MIND THAT BY THEN, COSMONAUT GHERMAN TITOV HAD ALREADY SPENT A DAY IN ORBIT, CIRCLING THE EARTH 17 TIMES.

AFTER FIVE YEARS AND MORE THAN A BILLION DOLLARS,

THE UNITED STATES WAS STILL LOSING THE SPACE RACE.

IS THERE ANY OTHER SPACE PROGRAM WHICH PROMISES DRAMATIC RESULTS IN WHICH WE COULD WIN?

SEPTEMBER 1962

RICE UNIVERSITY

HOUSTON, TEXAS

THE EXPLORATION OF SPACE WILL GO AHEAD, WHETHER WE JOIN IN IT OR NOT...

...WE MEAN TO BE A PART OF IT--WE MEAN TO LEAD IT.

EARLIER THAT SUMMER, IN WASHINGTON, D.C., DEKE ARRIVED IN A ROOM FILLED WITH DOCTORS.

THEY TOOK TURNS LISTENING TO HIS HEARTBEAT.

FOR THE EYES OF THE WORLD NOW LOOK INTO SPACE, TO THE MOON AND TO THE PLANETS BEYOND,

AND WE HAVE VOWED THAT WE SHALL NOT SEE IT GOVERNED BY A HOSTILE FLAG OF CONQUEST, BUT BY A BANNER OF FREEDOM AND PEACE...

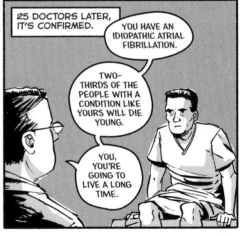

25 DOCTORS LATER, IT'S CONFIRMED.

YOU HAVE AN IDIOPATHIC ATRIAL FIBRILLATION.

TWO-THIRDS OF THE PEOPLE WITH A CONDITION LIKE YOURS WILL DIE YOUNG.

YOU, YOU'RE GOING TO LIVE A LONG TIME.

WE CHOOSE TO GO TO THE MOON.

"WE CHOOSE TO GO TO THE MOON IN THIS DECADE AND DO THE OTHER THINGS, NOT BECAUSE THEY ARE EASY, BUT BECAUSE THEY ARE HARD..."

"...BECAUSE THAT CHALLENGE IS ONE THAT WE ARE WILLING TO ACCEPT, ONE WE ARE UNWILLING TO POSTPONE, AND ONE WHICH WE INTEND TO WIN..."

HE WAS FINE.

SOMETHING WAS WRONG WITH HIS HEART, BUT DEKE WOULD BE FINE.

AND YET HE WASN'T FINE.
SOMETHING WAS WRONG WITH HIS HEART, AND THERE WAS NO WAY NASA WOULD LET HIM FLY.

MANY YEARS AGO THE GREAT BRITISH EXPLORER GEORGE MALLORY, WHO WAS TO DIE ON MOUNT EVEREST, WAS ASKED WHY DID HE WANT TO CLIMB IT.

"HE SAID, 'BECAUSE IT IS THERE.'"

WELL, SPACE IS THERE, AND WE'RE GOING TO CLIMB IT, AND THE MOON AND THE PLANETS ARE THERE, AND NEW HOPES FOR KNOWLEDGE AND PEACE ARE THERE.

"AND THEREFORE, AS WE SET SAIL, WE ASK GOD'S BLESSING ON THE MOST HAZARDOUS AND DANGEROUS AND GREATEST ADVENTURE ON WHICH MAN HAS EVER EMBARKED."

WE WERE GOING TO THE MOON. BUT DEKE WAS STAYING HERE.

KENNEDY'S CALL FOR A MOONSHOT OPENED A FLOODGATE OF FUNDING TO THE SPACE AGENCY.

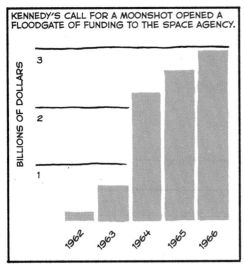

BILLIONS OF DOLLARS

3

2

1

1962 1963 1964 1965 1966

IT WAS ONE THING TO SEND A MAN INTO SPACE FOR MINUTES OR HOURS.

≈ 150 MILES

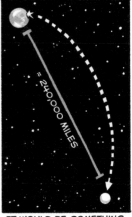

≈ 240,000 MILES

IT WOULD BE SOMETHING ELSE ENTIRELY TO PULL OFF WHAT PRESIDENT KENNEDY HAD PROMISED.

IN A WAY, NASA HAD TO START FROM SCRATCH.

A NEW ROCKET--THE SATURN V--THE BIGGEST EVER BUILT.

A NEW SPACECRAFT (OR TWO, AS IT TURNED OUT).

NEW LAUNCHPADS AND INFRASTRUCTURE.

NEW EMPLOYEES, NEW POSITIONS, NEW TITLES.

LOCKHEED

IBM

NASA

NEW BIDS, NEW CONTRACTORS.

R

W

GM
GENERAL MOTORS

AND, OF COURSE, NEW ASTRONAUTS.

DEKE WAS GROUNDED. BUT NASA WASN'T DONE WITH HIM.

IN THE BALLOONING BUREAUCRACY OF NASA, SOMEONE NEEDED TO SPEAK FOR THE ASTRONAUTS, AND THEY FIGURED IT WAS BETTER TO HAVE ONE OF THEIR OWN.

WE AIN'T GONNA HAVE SOME OUTSIDE WEENIE COMING IN, TELLING US WHAT TO DO.

SO IN LATE 1962, NASA MADE A NEW POSITION, JUST FOR DEKE.

HOW DOES "CHIEF ASTRONAUT" SOUND?

FROM THEN ON, IT WAS HIS JOB TO DECIDE WHO FLEW WHAT MISSION,

TO CORRAL THE GROWING ROSTER OF ASTRONAUTS,

AND TO RUN INTERFERENCE DURING THE INEVITABLE HEAD-BUTTING BETWEEN PILOT AND PENCIL PUSHER.

THE GLAMOUR OF HIS LIFE AS A MERCURY ASTRONAUT WAS GONE.

BUT NOW DEKE HAD SOMETHING ELSE,

A CALLING,

A MISSION,

A JOB:

FIND THE MEN WHO WOULD GO TO THE MOON.

CHAPTER XIII
EAGLE

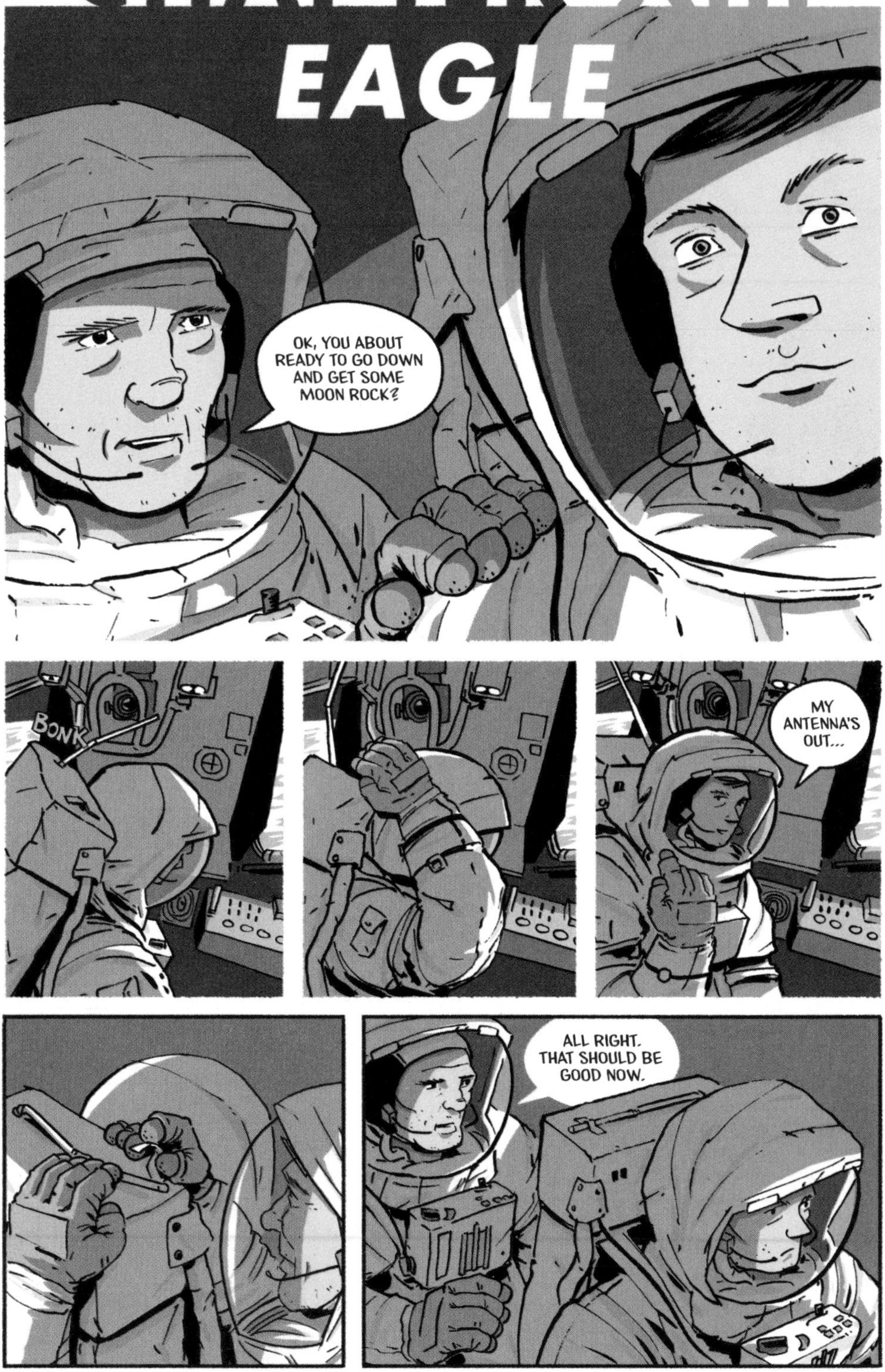
OK, YOU ABOUT READY TO GO DOWN AND GET SOME MOON ROCK?
BONK
MY ANTENNA'S OUT...
ALL RIGHT. THAT SHOULD BE GOOD NOW.

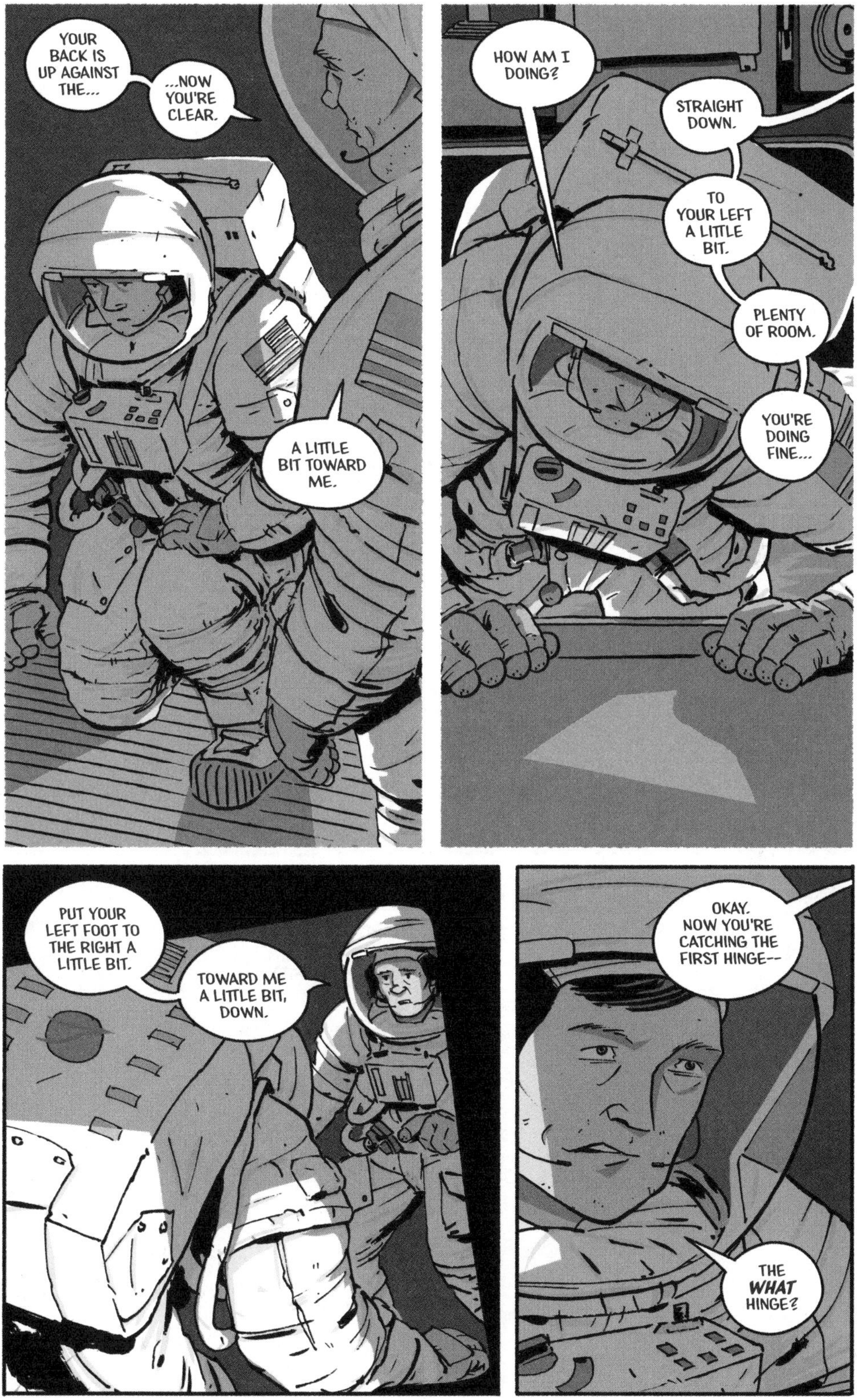
YOUR BACK IS UP AGAINST THE...
...NOW YOU'RE CLEAR.
A LITTLE BIT TOWARD ME.
HOW AM I DOING?
STRAIGHT DOWN.
TO YOUR LEFT A LITTLE BIT.
PLENTY OF ROOM.
YOU'RE DOING FINE...
PUT YOUR LEFT FOOT TO THE RIGHT A LITTLE BIT.
TOWARD ME A LITTLE BIT, DOWN.
OKAY. NOW YOU'RE CATCHING THE FIRST HINGE--
THE *WHAT* HINGE?

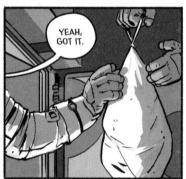

BEFORE HE COULD WALK ON THE MOON, ARMSTRONG HAD TO FIRST TAKE OUT THE TRASH.

OK, EVERYTHING'S NICE AND STRAIGHT IN HERE.

DID YOU GET THE MESA OUT?

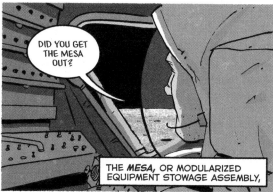

THE **MESA**, OR MODULARIZED EQUIPMENT STOWAGE ASSEMBLY,

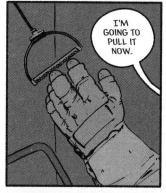

I'M GOING TO PULL IT NOW.

WAS A TRAPDOOR ON THE OUTSIDE OF THE *EAGLE*,

FULL OF TOOLS FOR THE ASTRONAUTS TO USE ON THEIR MOONWALK.

HAMMERS, SCOOPS, CONTAINERS FOR SOIL SAMPLES...

...AND, MOST IMPORTANT FOR EVERYONE WAITING BREATHLESSLY BACK ON EARTH,

A TELEVISION CAMERA.

HOUSTON. ROGER. WE COPY.

STANDING BY FOR YOUR TV.

CHAPTER XIV
CONTROL

WHEN YOU CONSIDER THE VELOCITIES AND FORCES
INVOLVED IN MISSILE LAUNCHINGS, YOU COME TO
REALIZE THAT HUMAN INTERVENTION IS NOT ONLY
IMPOSSIBLE FROM THE PHYSICAL STANDPOINT: IT IS
ACTUALLY UNDESIRABLE.

--WERNHER VON BRAUN

CONTROL OF SPACE MEANS CONTROL OF THE WORLD.

--LYNDON B. JOHNSON

MARGARET HAMILTON WORKED ALL DAY WHILE HER DAUGHTER WAS IN SCHOOL,

AND THEN AT NIGHT, AFTER DINNER, SHE WOULD WORK SOME MORE.

IT WAS 1965, AND NASA'S DEADLINE TO PUT A MAN ON THE MOON WAS HALFWAY UP.

DON'T TOUCH ANYTHING ELSE.

BUT YOU CAN PLAY WITH THIS.

IT'S THE SAME COMPUTER THAT THE ASTRONAUTS ARE GOING TO USE WHEN THEY GO TO THE MOON.

HER JOB WAS TO INSPECT THE PROGRAMMING ON THE APOLLO GUIDANCE COMPUTER FOR BUGS.

TEDIOUS NIGHTS OF TRIAL AND ERROR,

CLICK CLACK

PUNCTUATED EVERY NOW AND THEN BY A BREAKTHROUGH.

MOMMM... I THINK IT'S BROKEN.

WHAT HAPPENED?

NOTHING, I SWEAR.

I JUST PRESSED A BUTTON...

NO, HONEY, THIS IS GOOD. SHOW ME WHAT YOU DID.

HAMILTON WAS 29 YEARS OLD AND ONE OF THE FEW ENGINEERS WORKING FOR NASA WHO WASN'T A MAN.

MY FIRST ASSIGNMENT WAS LITERALLY CALLED "FORGET IT."

IT WAS A PIECE OF SOFTWARE THAT ONLY KICKED IN IN CASE OF A LAUNCH ABORT,

WHICH HER COLLEAGUES ASSUMED WOULD NEVER BE NEEDED.

BUT MARGARET HAMILTON HAD A KNACK FOR ANTICIPATING THE UNEXPECTED.

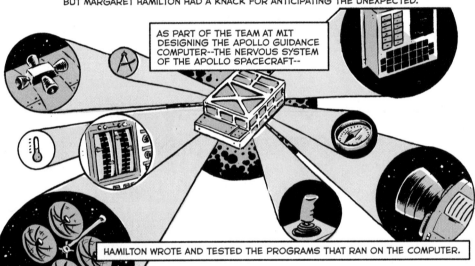

AS PART OF THE TEAM AT MIT DESIGNING THE APOLLO GUIDANCE COMPUTER--THE NERVOUS SYSTEM OF THE APOLLO SPACECRAFT--

HAMILTON WROTE AND TESTED THE PROGRAMS THAT RAN ON THE COMPUTER.

SHE STARTED CALLING WHAT SHE WAS DOING "SOFTWARE ENGINEERING,"

WHICH MADE IT SOUND MORE LEGITIMATE.

MOST OF THE MEN SHE WORKED WITH THOUGHT IT WAS A JOKE.

TO THEM, ENGINEERING MEANT DESIGNING SOMETHING TANGIBLE, HARDWARE.

PROGRAMMING--WRITING SOFTWARE-- HAD ALWAYS BEEN WOMEN'S WORK.

IN THE EARLY 20TH CENTURY, THE WORD "COMPUTER" REFERRED NOT TO A MACHINE BUT TO A JOB DESCRIPTION,

TO A PERSON WITH AN APTITUDE FOR MATH WHO COULD PERFORM CALCULATIONS BY HAND.

JOBS

"COMPUTERS DO NOT NEED ADVANCED DEGREES"

THAT WAS CODE FOR "WOMEN SHOULD APPLY."

DURING WORLD WAR II AND IN THE SUBSEQUENT AEROSPACE BOOM OF THE 1950s,

THOUSANDS OF FEMALE COMPUTERS ENTERED THE WORKFORCE.

EVEN MINOR CHALLENGES IN ROCKET ENGINEERING REQUIRED STAGGERING AMOUNTS OF MATHEMATICAL COMPUTATION.

TO KEEP UP WITH ALL OF THAT WORK, ENGINEERS RELIED ON LEGIONS OF COMPUTERS TO PROCESS THE DATA.

AS HELEN LING, HEAD OF THE COMPUTING DEPARTMENT AT THE JET PROPULSION LABORATORY, PUT IT:

ENGINEERS MAKE UP THE PROBLEMS AND WE SOLVE THEM.

AND YET, BECAUSE COMPUTERS WERE MOSTLY WOMEN,

AND BECAUSE ROCKETS BELONGED--AT LEAST IN THE PUBLIC IMAGINATION--TO THE REALM OF SWAGGERING ENGINEERS LIKE WERNHER VON BRAUN,

THEIR WORK, THOUGH INTEGRAL TO THE SPACE RACE, WAS NEVER IN THE SPOTLIGHT.

EVEN FURTHER FROM THE SPOTLIGHT WERE THE AFRICAN-AMERICAN COMPUTERS,

WOMEN LIKE KATHERINE JOHNSON AND MARY JACKSON,

PROMINENT MEMBERS OF THE SEGREGATED COMPUTING DIVISION AT NASA'S LANGLEY RESEARCH CENTER IN ALABAMA.

IF IT SEEMS AS IF THE STORY OF NASA IN THE APOLLO ERA IS ALL BUT ENTIRELY THE STORY OF WHITE MEN, THERE ARE TWO REASONS FOR THAT:

FIRST, THE EVENTS AND FIGURES DOCUMENTED BY NASA AND AMPLIFIED BY THE PRESS TENDED TO EXCLUDE THE CONTRIBUTIONS OF WOMEN AND PEOPLE OF COLOR.

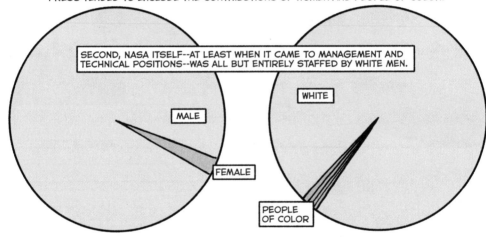

SECOND, NASA ITSELF--AT LEAST WHEN IT CAME TO MANAGEMENT AND TECHNICAL POSITIONS--WAS ALL BUT ENTIRELY STAFFED BY WHITE MEN.

MALE

FEMALE

WHITE

PEOPLE OF COLOR

ALTHOUGH THE APOLLO PROGRAM COINCIDED WITH THE CIVIL RIGHTS MOVEMENT, NASA WAS SLOWER THAN THE REST OF THE COUNTRY TO INTEGRATE.

3/4 OF NASA'S FACILITIES WERE LOCATED IN THE SOUTH,

AND NASA'S AEROSPACE CONTRACTORS HAD THEIR OWN INTERNAL LEGACIES OF DISCRIMINATION.

IN THE WORDS OF ONE HISTORIAN: "AFRICAN AMERICANS PUSHED BROOMS; WHITES BUILT AIRPLANES."

DOROTHY VAUGHAN, A SUPERVISOR OF THE LANGLEY COMPUTERS, WAS CIRCUMSPECT.

I CHANGED WHAT I COULD, AND WHAT I COULDN'T, I ENDURED.

WHEN VAUGHAN STARTED AT LANGLEY DURING WORLD WAR II,

THE TOOLS OF HER TRADE WERE A MECHANICAL CALCULATOR,

BOOKS OF NUMBER TABLES,

A SLIDE RULE,

AND A PENCIL.

BY THE EARLY 1960S, THOSE TOOLS HAD BEEN REPLACED BY AN ELECTRIC COMPUTER.

VAUGHAN HAD ANTICIPATED THIS SHIFT.

SHE LEARNED TO ADAPT HER FLUENCY IN MATH TO THE NEW LANGUAGE OF PROGRAMMING.

AS MACHINES TOOK OVER THE WORK OF FEMALE COMPUTERS,

MEN BEGAN TO TAKE OVER THE FIELD OF COMPUTING.

BY THE TIME NASA WAS READY TO SEND A MAN TO THE MOON,

THE WORD "COMPUTER" HAD SHED ALL SEMBLANCE OF ITS FORMER MEANING.

153

WHEN HER DAUGHTER INADVERTENTLY CRASHED THE APOLLO GUIDANCE COMPUTER, IT WAS A REVELATION TO MARGARET HAMILTON.

THE COMPUTER WAS DESIGNED SO THAT AT EACH PHASE OF A MISSION IT COULD RUN SEVERAL PROGRAMS AT ONCE,

GIVING PRIORITY TO WHICHEVER OPERATION WAS MOST IMPORTANT IN THE MOMENT.

HAMILTON EXPLAINED THE PROBLEM TO HER BOSS.

IF ANYONE TELLS THE COMPUTER TO RUN PROGRAMS FROM DIFFERENT PHASES OF THE MISSION, IT CRASHES.

ALL YOU'D NEED TO DO IS KEY IN THE WRONG CODE--

LET ME STOP YOU RIGHT THERE.

ASTRONAUTS DON'T MAKE MISTAKES.

HAMILTON'S MODEST SUGGESTION--

THAT HER TEAM PROGRAM INTO THE COMPUTER SOME WAY TO DETECT AND RECOVER FROM ERRORS--

HINTED AT A LARGER DEBATE AT NASA ABOUT HOW MUCH CONTROL APOLLO ASTRONAUTS WOULD HAVE ON THEIR VOYAGE TO THE MOON.

WOULD THEY BE PILOTS OR PASSENGERS?

ENGINEERS LIKE VON BRAUN ENVISIONED FULLY AUTOMATED SPACECRAFT.

WE LIKE TO THINK OF MAN AS AN AMAZINGLY VERSATILE COMPUTER.

BUT IN MISSILE TERMS, HE IS OUTRAGEOUSLY SLOW AND CUMBERSOME.

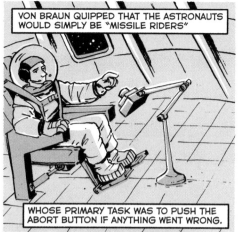

VON BRAUN QUIPPED THAT THE ASTRONAUTS WOULD SIMPLY BE "MISSILE RIDERS"

WHOSE PRIMARY TASK WAS TO PUSH THE ABORT BUTTON IF ANYTHING WENT WRONG.

OF COURSE, THE ASTRONAUTS, ALL OF THEM PILOTS, REFUSED TO BE TREATED--AS THEY PUT IT--LIKE "SPAM IN A CAN."

BUT WHAT IF THERE WERE NO ASTRONAUTS?

THAT WOULD MEAN NO LIFE-SUPPORT SYSTEMS,

NO CREW CABIN,

NO INSTRUMENT PANELS,

NO MORE NEED FOR A RETURN FLIGHT,

WHICH WOULD MEAN LESS PROPELLANT AND MORE ROOM FOR SCIENTIFIC EQUIPMENT.

AND WITHOUT A CREW TO KEEP ALIVE, WHY NOT RANGE BEYOND THE MOON, INTO THE SOLAR SYSTEM, OR FARTHER STILL?

TAKE OUT THE ASTRONAUTS AND THE POSSIBILITIES OF SPACEFLIGHT EXPAND.

THIS WAS THE VISION THAT EMERGED FROM THE JET PROPULSION LABORATORY IN CALIFORNIA,

THE RESEARCH CENTER THAT DESIGNED NASA'S SPACE PROBES.

JPL

JET PROPULSION LABORATORY

IN 1960--BEFORE APOLLO, BEFORE MERCURY--THERE WAS SURVEYOR, JPL'S SCIENTIFIC EXPEDITION TO THE MOON. THE SURVEYOR PROGRAM WOULD USE TWO KINDS OF ROBOT:

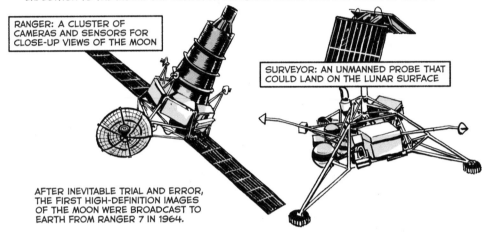

RANGER: A CLUSTER OF CAMERAS AND SENSORS FOR CLOSE-UP VIEWS OF THE MOON

SURVEYOR: AN UNMANNED PROBE THAT COULD LAND ON THE LUNAR SURFACE

AFTER INEVITABLE TRIAL AND ERROR, THE FIRST HIGH-DEFINITION IMAGES OF THE MOON WERE BROADCAST TO EARTH FROM RANGER 7 IN 1964.

AND THEN TWO YEARS LATER, IN JUNE OF 1966, SURVEYOR 1 LANDED IN THE OCEAN OF STORMS...

...JUST TO THE EAST OF THE SITE WHERE, SIX MONTHS EARLIER, SERGEI KOROLEV'S LUNA 9 HAD LANDED FIRST. ONCE AGAIN, NASA WAS RUNNER-UP.

REGARDLESS, HERE IT WAS, THE AMERICAN MOON LANDING!

THE RACE WAS OVER!

THE U.S. HAD COME IN SECOND, BUT THE KNOWLEDGE GAINED WOULD BROADEN OUR UNDERSTANDING OF HUMANITY'S PLACE IN THE COSMOS!

THE RACE TO THE MOON WAS OVER.

SO WHY WAS NASA STILL RUNNING?

BECAUSE, AS JAMES WEBB--THE NASA ADMINISTRATOR WHO LED THE APOLLO PROGRAM--ONCE ARGUED:

IT IS MAN, AND NOT MERELY MACHINES, IN SPACE THAT CAPTURES THE IMAGINATION OF THE WORLD.

ROBOTS ON THE MOON WERE NO SUBSITUTE FOR HUMAN FOOTPRINTS.

GOING TO THE MOON WAS NEVER ABOUT THE MOON ITSELF.

GOING TO THE MOON WAS A QUEST,

AND LIKE ANY QUEST,

THE QUARRY WAS LARGELY SYMBOLIC.

JOHN F. KENNEDY UNDERSTOOD THIS.

I'M NOT THAT INTERESTED IN SPACE.

I THINK IT'S GOOD; I THINK WE OUGHT TO KNOW ABOUT IT.

BUT WE'RE TALKING ABOUT THESE *FANTASTIC* EXPENDITURES

AND THE ONLY JUSTIFICATION FOR IT IS BECAUSE WE HOPE TO BEAT **THEM**.

THE SOVIET UNION MADE THIS A TEST

SO THAT'S WHY WE'RE DOING IT.

THE MACHINES THEMSELVES, AND THE MEN WHO WOULD FLY THEM, WERE MERELY MEANS TO AN END.

IN 1962, WHEN CONGRESS FUNDED PRESIDENT KENNEDY'S PROMISE TO LAND A MAN ON THE MOON,

ASTRONAUTS, AND ALL THEY HAD COME TO REPRESENT, BECAME NASA'S TOP PRIORITY.

JOHN GLENN SUMMED UP THE MOOD OF HIS FELLOW ASTRONAUTS:

NOW WE CAN GET RID OF SOME OF THAT AUTOMATIC EQUIPMENT AND LET MAN TAKE OVER.

NEVERTHELESS, IN THE GULF BETWEEN JOHN GLENN'S FIRST ORBIT AND THE MOON LANDING TO COME, THERE LOOMED A LONG LIST OF UNKNOWNS.

THE ATTEMPT TO SHORTEN THAT LIST WAS CALLED PROJECT GEMINI:

NASA

GEMINI

TEN PAIRS OF ASTRONAUTS, LAUNCHED INTO ORBIT EVERY TWO MONTHS, STARTING IN 1965.

THE GEMINI FLIGHTS TACKLED ALL THE ESSENTIAL CHALLENGES OF A MOONSHOT.

CAN AN ASTRONAUT'S SUIT KEEP HIM ALIVE IN THE OPEN ABYSS OF SPACE?

I'M SEPARATING FROM THE SPACECRAFT...

OOPS! THERE GOES YOUR GLOVE!

WHAT'S THE LONGEST MISSION THAT A SHIP AND ITS CREW CAN ENDURE?

FOURTEEN DAYS...NOBODY EVER SAID IT WOULD BE EASY, I GUESS.

WAS MANEUVERING A SPACECRAFT SIMILAR TO FLYING AN AIRPLANE?

I'VE BEEN STRUGGLING HERE TO NOT LET THE THING GET TOO FAR FROM ME.

WHAT HAPPENS WHEN TWO SPACECRAFT TRY TO NAVIGATE THE TRICKY PHYSICS OF ORBIT?

...WE'RE ALL SITTING UP HERE PLAYING BRIDGE TOGETHER.

THIS LAST TASK WAS THE BIGGEST UNKNOWN.

A CRUCIAL STEP ON THE WAY TO THE MOON REQUIRED ASTRONAUTS TO PERFORM COMPLEX MANEUVERS WITH MULTIPLE SPACECRAFT.

WHICH, FOR A BUNCH OF SEASONED TEST PILOTS, DIDN'T SEEM LIKE A VERY BIG DEAL.

159

WHAT THOSE FIGHTER JOCKS ARE GONNA HAVE TO LEARN IS THAT "FLYING" IN SPACE IS A WHOLE DIFFERENT BALL GAME.

JOHN S. LLEWELLYN--A FORMER MARINE AND EARLY MEMBER OF MISSION CONTROL--WAS WELL-VERSED IN THE THEORETICAL CHALLENGES OF SPACEFLIGHT.

SAY THIS PEN IS YOUR SPACESHIP,

AND SAY THIS MATCHBOOK IS ANOTHER.

AND YOU WANT THEM BOTH TO MEET UP IN ORBIT.

A FIGHTER PILOT WOULD JUST THROTTLE UP UNTIL HE'S GOING FAST ENOUGH TO OVERTAKE THE MATCHBOOK.

BUT IN ORBIT, WITH A MOVE LIKE THAT YOU'LL ACTUALLY END UP FARTHER AWAY THAN WHEN YOU STARTED.

THAT'S BECAUSE OF HOW GRAVITY WORKS IN SPACE.

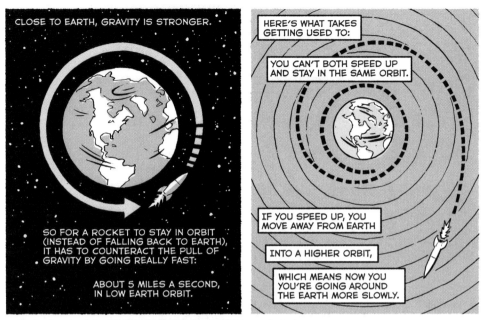

CLOSE TO EARTH, GRAVITY IS STRONGER.

SO FOR A ROCKET TO STAY IN ORBIT (INSTEAD OF FALLING BACK TO EARTH), IT HAS TO COUNTERACT THE PULL OF GRAVITY BY GOING REALLY FAST:

ABOUT 5 MILES A SECOND, IN LOW EARTH ORBIT.

HERE'S WHAT TAKES GETTING USED TO:

YOU CAN'T BOTH SPEED UP AND STAY IN THE SAME ORBIT.

IF YOU SPEED UP, YOU MOVE AWAY FROM EARTH

INTO A HIGHER ORBIT,

WHICH MEANS NOW YOU YOU'RE GOING AROUND THE EARTH MORE SLOWLY.

TO CATCH UP WITH THE MATCHBOOK, THE PEN, COUNTERINTUITIVELY, HAS TO LOSE SOME SPEED.

BY ACCELERATING IN THE OPPOSITE DIRECTION,

THE PEN FALLS DOWN TO A LOWER ORBIT,

WHICH MEANS IT'S NOW ACTUALLY GOING FASTER THAN THE MATCHBOOK.

THEN IT'S JUST A MATTER OF JUMPING OFF THE MERRY-GO-ROUND AT THE RIGHT TIME,

BY ACCELERATING BACK INTO THE HIGHER ORBIT NEXT TO THE MATCHBOOK.

AND THAT, KIDS, IS WHAT WE CALL A SPACE RENDEZVOUS.

TO PULL OFF THESE COMPLEX ORBITAL MANEUVERS, ASTRONAUTS NEEDED MORE THAN INSTINCT OR SKILL.

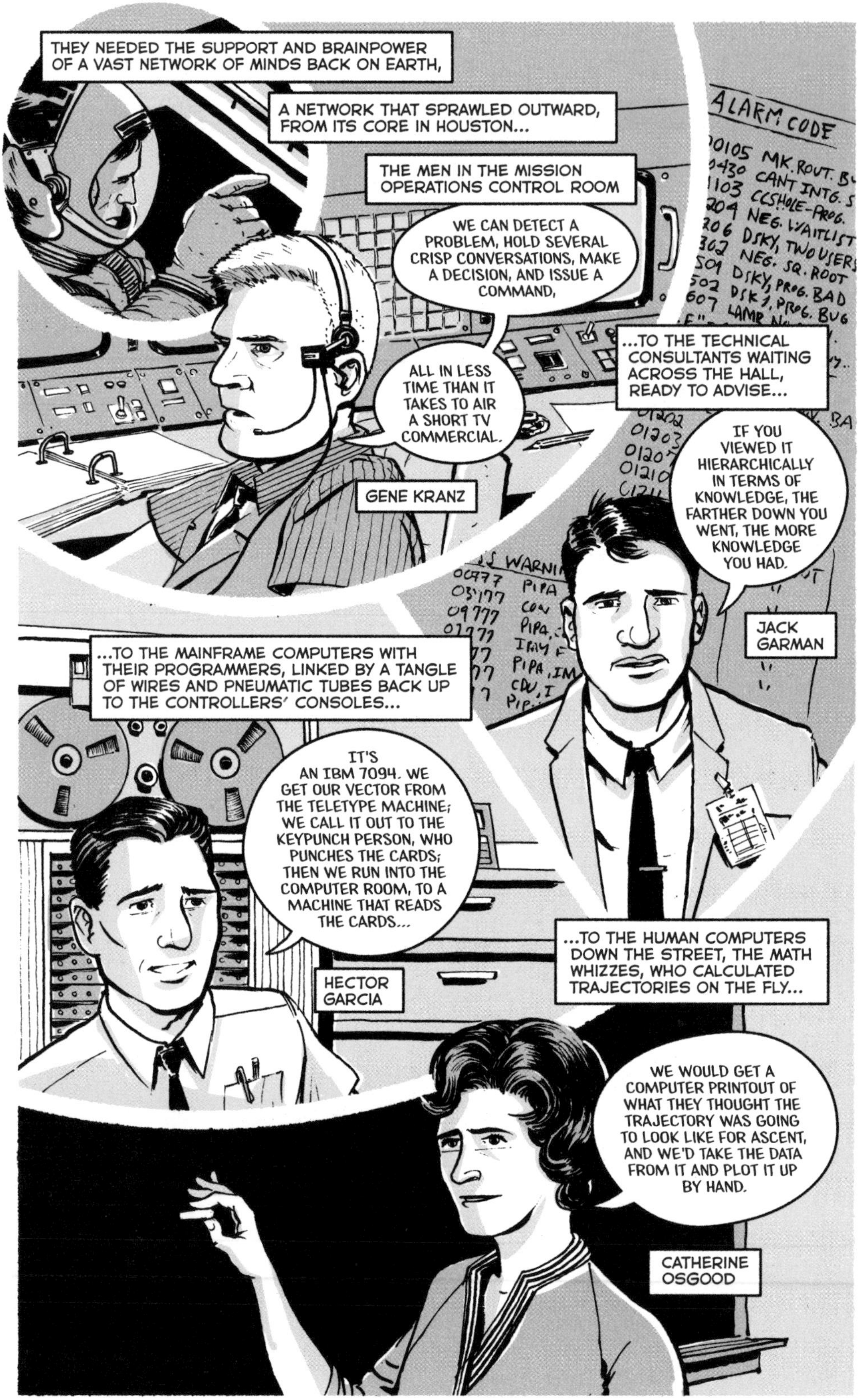
THEY NEEDED THE SUPPORT AND BRAINPOWER OF A VAST NETWORK OF MINDS BACK ON EARTH,
A NETWORK THAT SPRAWLED OUTWARD, FROM ITS CORE IN HOUSTON...
THE MEN IN THE MISSION OPERATIONS CONTROL ROOM
ALARM CODE
WE CAN DETECT A PROBLEM, HOLD SEVERAL CRISP CONVERSATIONS, MAKE A DECISION, AND ISSUE A COMMAND,
ALL IN LESS TIME THAN IT TAKES TO AIR A SHORT TV COMMERCIAL.
GENE KRANZ
...TO THE TECHNICAL CONSULTANTS WAITING ACROSS THE HALL, READY TO ADVISE...
IF YOU VIEWED IT HIERARCHICALLY IN TERMS OF KNOWLEDGE, THE FARTHER DOWN YOU WENT, THE MORE KNOWLEDGE YOU HAD.
JACK GARMAN
...TO THE MAINFRAME COMPUTERS WITH THEIR PROGRAMMERS, LINKED BY A TANGLE OF WIRES AND PNEUMATIC TUBES BACK UP TO THE CONTROLLERS' CONSOLES...
IT'S AN IBM 7094. WE GET OUR VECTOR FROM THE TELETYPE MACHINE; WE CALL IT OUT TO THE KEYPUNCH PERSON, WHO PUNCHES THE CARDS; THEN WE RUN INTO THE COMPUTER ROOM, TO A MACHINE THAT READS THE CARDS...
HECTOR GARCIA
...TO THE HUMAN COMPUTERS DOWN THE STREET, THE MATH WHIZZES, WHO CALCULATED TRAJECTORIES ON THE FLY...
WE WOULD GET A COMPUTER PRINTOUT OF WHAT THEY THOUGHT THE TRAJECTORY WAS GOING TO LOOK LIKE FOR ASCENT, AND WE'D TAKE THE DATA FROM IT AND PLOT IT UP BY HAND.
CATHERINE OSGOOD

...TO THE GOVERNMENT RESEARCH FACILITIES ACROSS THE COUNTRY...

AMES RESEARCH CENTER, MOUNTAINVIEW, CALIFORNIA.

LEWIS RESEARCH CENTER, COLUMBUS, OHIO.

LANGLEY RESEARCH CENTER, HAMPTON, VIRGINIA.

MARSHALL SPACEFLIGHT CENTER, HUNTSVILLE, ALABAMA.

...TO THE DESIGN SHOPS AND FACTORIES OF THE COMPANIES CONTRACTED BY NASA, WHERE, DURING MISSIONS, WORKERS WAITED AT THE READY WITH BLUEPRINTS AND MANUALS...

...AND FINALLY TO THE FAR-FLUNG OPERATORS AT TRACKING STATIONS AROUND THE WORLD, THEIR ANTENNAS TUNED SKYWARD,

KEEPING CONTACT WITH THE ASTRONAUTS NO MATTER WHERE THEY WERE IN ORBIT.

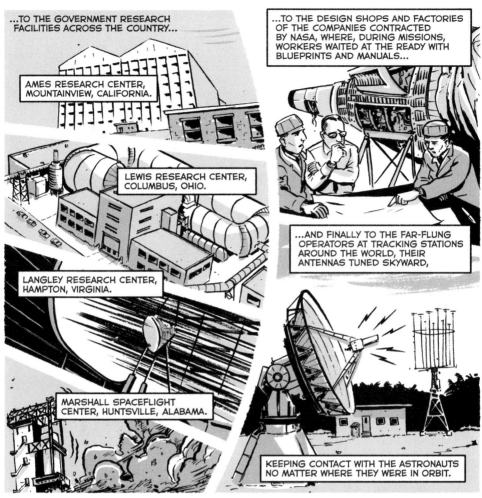

ALL TOLD, WITH SOME 400,000 PEOPLE WORKING TO GET AMERICAN ASTRONAUTS TO THE MOON, IN WHAT WAY DOES IT EVEN MAKE SENSE TO ASK WHO WAS IN CONTROL?

WAS IT THE ASTRONAUTS,

THE PUBLIC FACES OF APOLLO, WHO PILOTED THEIR SPACECRAFT,

THOUGH NOT WITHOUT THE HELP OF COMPUTERS?

OR WAS IT THE GROUND CONTROLLERS IN HOUSTON,

MEN LIKE KRANZ, WHOSE JOB DESCRIPTION SIMPLY READ "MAY TAKE ANY ACTION NECESSARY FOR MISSION SUCCESS"?

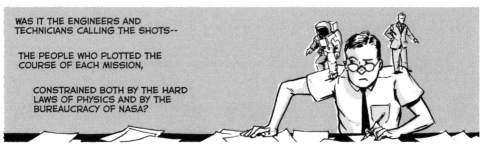

WAS IT THE ENGINEERS AND TECHNICIANS CALLING THE SHOTS--

THE PEOPLE WHO PLOTTED THE COURSE OF EACH MISSION,

CONSTRAINED BOTH BY THE HARD LAWS OF PHYSICS AND BY THE BUREAUCRACY OF NASA?

OR WAS IT THOSE ADMINISTRATORS, THE GOVERNMENT APPOINTEES WHO SET THE GOALPOSTS OF THE APOLLO PROGRAM, ITS AMBITIONS AND BUDGETS?

WAS IT THE THE POLITICIANS WHO APPOINTED NASA'S ADMINISTRATORS--

THREE DIFFERENT PRESIDENTS, MEMBERS OF CONGRESS--

WHO THEMSELVES WERE BEHOLDEN TO THE VOTES OF THEIR TAXPAYING CONSTITUENTS?

WERE THE AMERICAN PEOPLE IN CONTROL OF THIS NATIONAL PROJECT?

OR WAS THE COUNTRY TAKING ITS CUES FROM A TECHNOCRATIC ELITE,

FROM A MILITARY-INDUSTRIAL COMPLEX WHOSE QUEST FOR POWER AND PROFIT COMMANDEERED THE IMAGINATION OF A WHOLE SOCIETY?

OR MAYBE WAS THERE SOME DEEPER ATAVISTIC DREAM AT WORK,

SOME INEVITABLE HUMAN IMPULSE TO SEEK OUT WONDER?

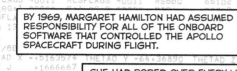

BY 1969, MARGARET HAMILTON HAD ASSUMED RESPONSIBILITY FOR ALL OF THE ONBOARD SOFTWARE THAT CONTROLLED THE APOLLO SPACECRAFT DURING FLIGHT.

SHE HAD PORED OVER EVERY LINE OF CODE, ROOTING OUT ERRORS,

BUILDING SAFEGUARDS SO THAT A MISTIMED KEYSTROKE COULDN'T TRIGGER A CATASTROPHE.

HER EXPERTISE MEANT THAT ON THE NIGHT OF THE MOON LANDING,

SHE WAS CLOISTERED IN A CLASSROOM AT MIT,

WAITING, ON CALL, IN CASE NASA NEEDED HER ADVICE.

FOR A NIGHT SHE WAS ONE OF THE ANOINTED FEW,

A SINGULAR, SPECIALIZED MIND AT THE CENTER OF THIS HISTORIC MOMENT.

·OPERNICVS· GALILEO·
·YCHO BRAHE· KEPLER·

·NEWTON·

AND FOR A NIGHT, SHE WAS JUST LIKE EVERYONE ELSE:

THOUSANDS OF MILES AWAY FROM THE ACTION,

ACUTELY AWARE THAT WHATEVER CAME NEXT WAS BEYOND HER CONTROL.

165

CHAPTER XV
HOUSTON

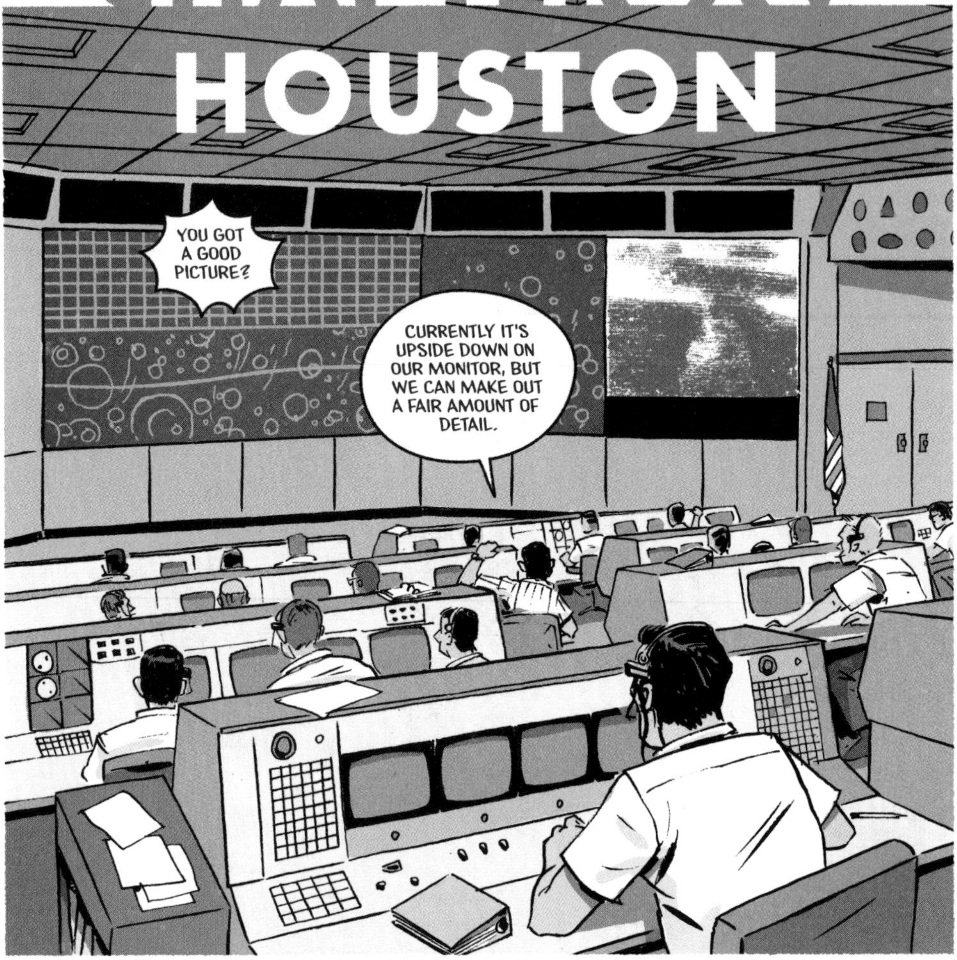
YOU GOT A GOOD PICTURE?
CURRENTLY IT'S UPSIDE DOWN ON OUR MONITOR, BUT WE CAN MAKE OUT A FAIR AMOUNT OF DETAIL.

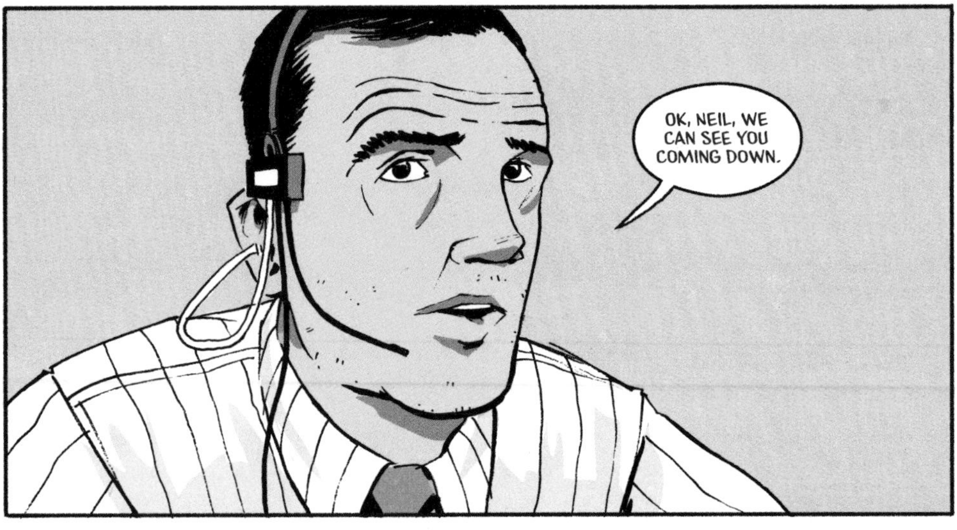
OK, NEIL, WE CAN SEE YOU COMING DOWN.

TO ANYONE WHO DIDN'T KNOW BETTER,

IT MIGHT SEEM ODD THAT SO MUCH OF THE WORLD ON JULY 20, 1969,

WAS WAITING FOR A MAN TO CLIMB DOWN A LADDER.

BUT THAT SUMMER SUNDAY WAS ONE OF THOSE RARE MOMENTS WHEN NORMAL PEOPLE GET TO STAKE A CLAIM ON THE SOMBER, STUFFY TIME LINE OF HISTORY.

I'M AT THE FOOT OF THE LADDER.

WHEN THOSE WHO LIVED THROUGH IT GET TO SAY "I REMEMBER WHERE I WAS WHEN..."

AND FOR A GENERATION MORE ACCUSTOMED TO USING THAT PHRASE IN THE FACE OF TRAGEDY,

THE CAREFUL, CONFIDENT MOVEMENTS OF NEIL ARMSTRONG MUST HAVE SEEMED LIKE A REASSURANCE.

THE LM FOOTPADS ARE ONLY DEPRESSED IN THE SURFACE ABOUT...

AS IF TO SAY, THIS, AT LEAST, IS GOOD.

...ABOUT ONE OR TWO INCHES...

...ALTHOUGH THE SURFACE APPEARS TO BE...

THIS WILL BE THE BEGINNING OF SOMETHING NEW--AND BETTER.

THE MOMENT WOULDN'T HAVE BEEN A MOMENT AT ALL,

...TO BE VERY, VERY FINE-GRAINED, AS YOU...

...AS YOU GET CLOSE TO IT.

IF IT WEREN'T BEING BROADCAST LIVE ON TELEVISION AND RADIO.

ANYONE WHO WANTED TO COULD FOLLOW ALONG,

IT'S ALMOST LIKE A POWDER.

OK. I'M GOING TO STEP OFF THE LM NOW...

AS IF BEARING WITNESS WERE ITSELF A KIND OF PARTICIPATION.

CHAPTER XVI
MOONBOUND

JULY 16, 1969

2:03 A.M.

I'M UP...

DEE O'HARA GAVE HERSELF AN EXTRA HOUR TO GET TO WORK THIS MORNING.

SHE'D BEEN AT THIS NURSING JOB LONG ENOUGH TO KNOW THAT TRAFFIC WOULD BE BAD.

ESPECIALLY TODAY.

AND SHE DIDN'T WANT TO THINK WHAT WOULD HAPPEN IF SHE WAS LATE.

AFTER ALL, IT'S NOT EVERY DAY THAT HER PATIENTS WERE ABOUT TO GO TO THE MOON.

COCOA BEACH, FLORIDA

T-MINUS 06H:59M:39S

YEARS AGO, THIS WAS A SLEEPY BEACH TOWN.

THEN CAME THE LAUNCHPADS NEXT DOOR AT CAPE CANAVERAL,

AND THEIR ROCKETS ROARING OUT OVER THE ATLANTIC.

NEXT WAS THE HANDFUL OF MEN WHO RODE THOSE ROCKETS--THE ASTRONAUTS.

AND FOLLOWING IN THEIR WAKE, THE PILGRIMS.

ANYONE AND EVERYONE WHO WANTED TO SEE WITH THEIR OWN EYES WHAT THE SPACE AGE LOOKED LIKE.

ON A LAUNCH DAY--

AND ON THIS LAUNCH DAY IN PARTICULAR--

EVERY ROOM WAS BOOKED IN THE ASTROCRAFT, THE POLARIS, AND THE SATELLITE MOTELS...

...THE BARS SERVED MOON LANDER COCKTAILS AND LIFTOFF MARTINIS UNTIL DAWN...

...BUT JUST PAST THE EDGE OF TOWN, THE REVELRY AND EXCITEMENT GAVE WAY TO A MORE RESERVED AND ANXIOUS ANTICIPATION.

AFTER SEVEN YEARS, BILLIONS OF DOLLARS, AND UNCOUNTABLE HOURS OF WORK, APOLLO 11 WAS READY.

THE TIME HAD COME.

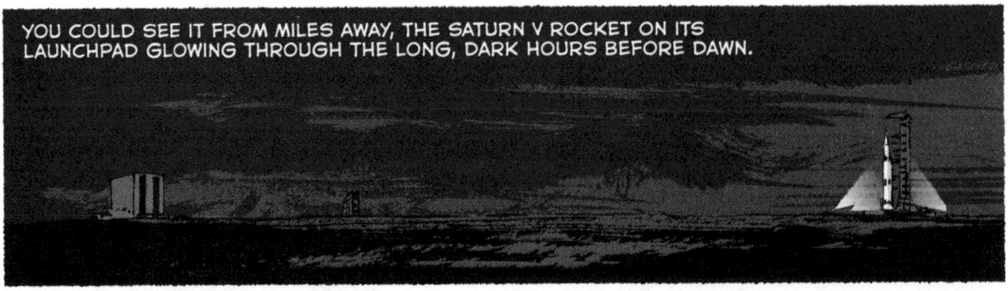
YOU COULD SEE IT FROM MILES AWAY, THE SATURN V ROCKET ON ITS LAUNCHPAD GLOWING THROUGH THE LONG, DARK HOURS BEFORE DAWN.

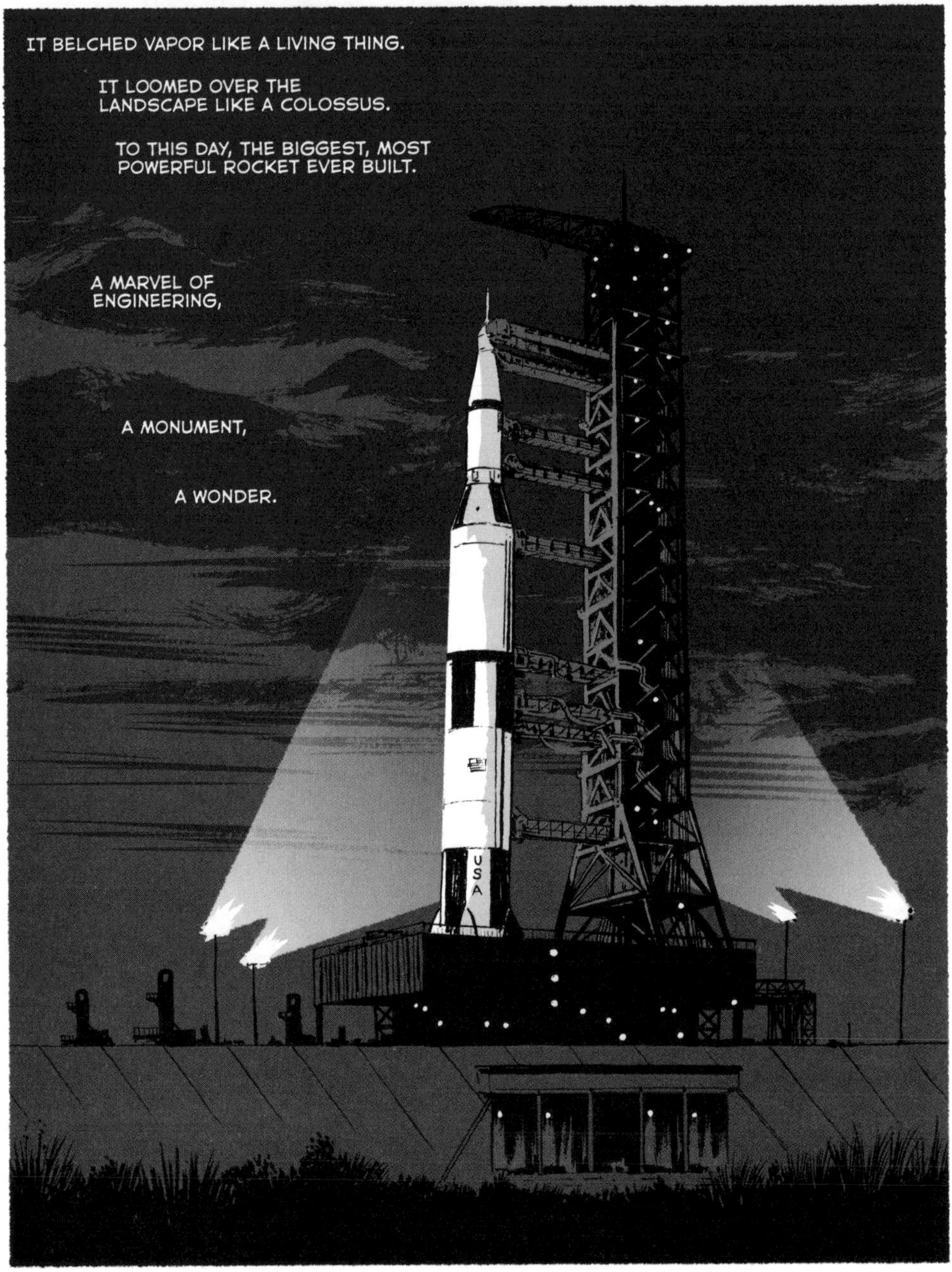
IT BELCHED VAPOR LIKE A LIVING THING.
IT LOOMED OVER THE LANDSCAPE LIKE A COLOSSUS.
TO THIS DAY, THE BIGGEST, MOST POWERFUL ROCKET EVER BUILT.
A MARVEL OF ENGINEERING,
A MONUMENT,
A WONDER.
USA

WHEN PRESIDENT KENNEDY COMMITTED THE COUNTRY TO LANDING A MAN ON THE MOON,

THE GENERAL ENTHUSIASM OF THE SPACE RACE OVERSHADOWED THE FACT THAT NO ONE WAS REALLY SURE HOW TO PROCEED.

OBVIOUSLY, NASA NEEDED A BIGGER ROCKET THAN THE ONE USED FOR PROJECT MERCURY, SEVEN YEARS EARLIER.

BUT HOW BIG?

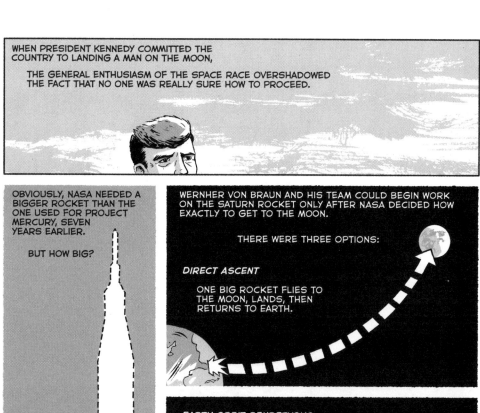

WERNHER VON BRAUN AND HIS TEAM COULD BEGIN WORK ON THE SATURN ROCKET ONLY AFTER NASA DECIDED HOW EXACTLY TO GET TO THE MOON.

THERE WERE THREE OPTIONS:

DIRECT ASCENT

ONE BIG ROCKET FLIES TO THE MOON, LANDS, THEN RETURNS TO EARTH.

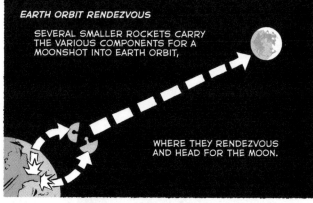

EARTH ORBIT RENDEZVOUS

SEVERAL SMALLER ROCKETS CARRY THE VARIOUS COMPONENTS FOR A MOONSHOT INTO EARTH ORBIT,

WHERE THEY RENDEZVOUS AND HEAD FOR THE MOON.

LUNAR ORBIT RENDEZVOUS

ONE ROCKET SENDS THE SPACECRAFT TO THE MOON,

WHERE IT SPLITS IN TWO,

WITH ONE CRAFT DESCENDING TO THE SURFACE AND THE OTHER WAITING IN ORBIT FOR THE RETURN FLIGHT.

THE DEBATE DOMINATED NASA IN THE FIRST YEARS OF THE APOLLO PROGRAM, FROM 1962 TO 1964. ULTIMATELY, LUNAR ORBIT RENDEZVOUS MADE THE MOST SENSE.

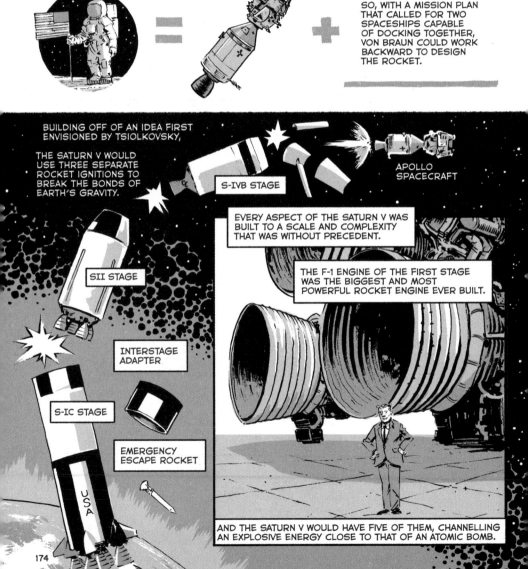

SO, WITH A MISSION PLAN THAT CALLED FOR TWO SPACESHIPS CAPABLE OF DOCKING TOGETHER, VON BRAUN COULD WORK BACKWARD TO DESIGN THE ROCKET.

BUILDING OFF OF AN IDEA FIRST ENVISIONED BY TSIOLKOVSKY,

THE SATURN V WOULD USE THREE SEPARATE ROCKET IGNITIONS TO BREAK THE BONDS OF EARTH'S GRAVITY.

S-IVB STAGE

APOLLO SPACECRAFT

EVERY ASPECT OF THE SATURN V WAS BUILT TO A SCALE AND COMPLEXITY THAT WAS WITHOUT PRECEDENT.

SII STAGE

THE F-1 ENGINE OF THE FIRST STAGE WAS THE BIGGEST AND MOST POWERFUL ROCKET ENGINE EVER BUILT.

INTERSTAGE ADAPTER

S-IC STAGE

EMERGENCY ESCAPE ROCKET

USA

AND THE SATURN V WOULD HAVE FIVE OF THEM, CHANNELLING AN EXPLOSIVE ENERGY CLOSE TO THAT OF AN ATOMIC BOMB.

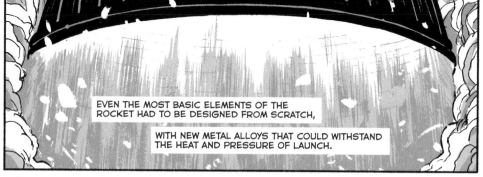

EVEN THE MOST BASIC ELEMENTS OF THE ROCKET HAD TO BE DESIGNED FROM SCRATCH,

WITH NEW METAL ALLOYS THAT COULD WITHSTAND THE HEAT AND PRESSURE OF LAUNCH.

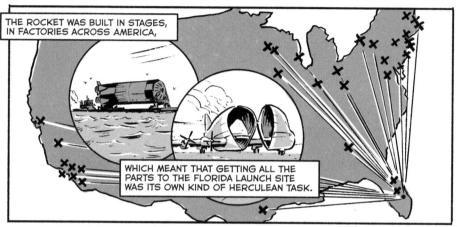

THE ROCKET WAS BUILT IN STAGES, IN FACTORIES ACROSS AMERICA,

WHICH MEANT THAT GETTING ALL THE PARTS TO THE FLORIDA LAUNCH SITE WAS ITS OWN KIND OF HERCULEAN TASK.

NEAR THE LAUNCH SITE, NASA BUILT THE VERTICAL ASSEMBLY BUILDING--AT THE TIME IT WAS THE LARGEST BUILDING IN THE WORLD, BIG ENOUGH TO HOUSE FOUR SATURN V ROCKETS.

INSIDE, THE STAGES OF THE ROCKET WERE HOISTED, STACKED, AND BOLTED TOGETHER.

ONCE ASSEMBLED, THE ROCKET INCHED TOWARD THE LAUNCH SITE ATOP A "CRAWLER TRANSPORTER," A SQUAT BEHEMOTH OF A VEHICLE, SO BIG IT NEEDED TWO DRIVERS, ONE AT EACH END.

THE SUPERLATIVES IN THIS STORY HAVE A WAY OF MASKING THE ALMOST CARTOONISH AUDACITY AT THE HEART OF IT ALL:

PUT THREE MEN ON TOP OF NEARLY SIX MILLION POUNDS OF EXPLOSIVES, AIM THEM AT THE MOON, AND LIGHT THE FUSE.

175

KENNEDY SPACE CENTER

T-MINUS 05H:41M:27S

MORNING, DEE.

GOOD MORNING, MR. SLAYTON.

WHY DON'T YOU GET YOURSELF SOME COFFEE.

I'LL GO WAKE UP THE BOYS.

SLAYTON HAD FLOWN TO THE CAPE TO SEND OFF HIS CREW.

AFTER ALL, HE WAS THE ONE WHO'D MADE THEM ASTRONAUTS IN THE FIRST PLACE,

AND IT WAS SLAYTON WHO, SEVEN MONTHS EARLIER, HAD CALLED THEM TO HIS OFFICE WITH NEWS THAT WOULD CHANGE THE TRAJECTORY OF THEIR LIVES.

I'LL CUT TO THE CHASE.

YOU'VE GOT THE APOLLO 11 FLIGHT.

YOU'RE IT, GUYS.

YOU GET FIRST CRACK AT LANDING ON THE MOON.

OF COURSE, THAT'S ASSUMING THE NEXT TWO APOLLO TEST FLIGHTS THIS SPRING ARE SUCCESSFUL...

RATHER THAN ASSIGN SPECIFIC MEN TO SPECIFIC MISSIONS, SLAYTON MAINTAINED A POOL OF ASTRONAUTS AND ROTATED THEM IN A SEQUENCE BASED ON SENIORITY.

WHEN IT WAS YOUR TURN, YOU FLEW WHATEVER MISSION WAS UP. AND A BACKUP CREW STOOD BY IN CASE ANYONE GOT SICK.

AND FOR ARMSTRONG, ALDRIN, AND COLLINS, IT WOULD SOON BE THEIR TURN.

PLEASE TAKE A NUMBER

THEY KNEW THEY WOULD FLY APOLLO 11, BUT ONLY TIME WOULD TELL WHETHER THAT WOULD BE THE MISSION TO LAND ON THE MOON.

DEKE, THANK YOU FOR HAVING CONFIDENCE IN US.

I FEEL DAMN FORTUNATE TO BE IN THIS ROOM RIGHT NOW.

WE WON'T LET YOU DOWN.

177

BEFORE BREAKFAST, THE CREW HAD TO GET ONE LAST PHYSICAL.

GOOD MORNING, NURSE O'HARA. WHAT ARE YOU DOING UP SO EARLY?

WELL, I DIDN'T HAVE ANYTHING ELSE TO DO...

AS IT WAS FOR EVERY SPACEFLIGHT SINCE ALAN SHEPARD'S FIRST MERCURY MISSION, BREAKFAST WAS STEAK AND EGGS.

WAIT, WHO'S GOING TO COOK FOR US UP ON THE MOON?

THEY DON'T PAY ME ENOUGH FOR THAT.

LEW HARTZELL HAD FIRST LEARNED TO COOK FOR HUNGRY MEN BACK IN THE MARINES.

OH, I PICKED UP A LITTLE FANCY STUFF LATER WHEN I WAS WORKING YACHTS.

THERE, YOU DON'T JUST SLICE A TOMATO, YOU MAKE IT LOOK PRETTY, LIKE A ROSE.

I DON'T DO THAT HERE. GIVE THESE BOYS AN HORS D'OEUVRE AND THEY'LL THROW IT AT YOU.

HARTZELL HAD BEEN THE ASTRONAUTS' COOK SINCE THE EARLY DAYS OF GEMINI...

...THOUGH HE TOOK A BREAK AFTER 1967.

THE YEAR OF THE FIRE.

JANUARY 27, 1967
APOLLO 1 WAS ON THE LAUNCHPAD, A MONTH BEFORE LIFTOFF.

HOW ARE WE GOING TO GET TO THE MOON IF WE CAN'T—

APOLLO 1, THIS IS STONY. HOW DO YOU READ?

ASTRONAUTS GUS GRISSOM, ED WHITE, AND ROGER CHAFFEE WERE ON BOARD FOR A PRACTICE COUNTDOWN.

IT WAS NOT GOING WELL.

JESUS CHRIST. I SAID, HOW ARE WE GOING TO GET TO THE MOON IF WE CAN'T TALK BETWEEN THREE BUILDINGS.

COMMS ON THE FRITZ, A STRANGE SMELL IN THE OXYGEN SUPPLY, AND A LONG LIST OF ELECTRICAL GLITCHES MEANT THAT THE FIRST MANNED APOLLO MISSION MIGHT BE DELAYED.

IT WAS A SLOW, TEDIOUS AFTERNOON.

SZZZKKKT...*FIRE*...SZZZZKKKT

DID YOU CATCH THAT?

SAY AGAIN, APOLLO 1?

SSSZZZKKKT... WE'VE GOT A *FIRE* IN THE COCKPIT. IT'S... SSZZZZKKKTTT

WE GOT A BAD FIRE...SSSZZZKKKTTT...*I'M GETTING OUT*...WE'RE BURNING...SSZZKKKTT

SLAYTON AND GRISSOM WERE CLOSE FRIENDS.

ONCE, GUS HAD SAVED DEKE FROM DROWNING.

BUT NOW THERE WAS NOTHING DEKE COULD DO.

TO AVOID DECOMPRESSION PROBLEMS, NASA'S SPACECRAFT WERE DESIGNED SO THAT ASTRONAUTS BREATHED PURE OXYGEN.

IT HAD BEEN THAT WAY SINCE PROJECT MERCURY.

DON'T TOUCH IT. IT'S TOO HOT.

WHAT NO ONE HAD CONSIDERED WAS THAT A COCKPIT PRESSURIZED WITH FLAMMABLE GAS COULD BE A DEATHTRAP.

ALL IT TOOK WAS A SPARK.

SEALED BEHIND A HATCH THAT TOOK AT BEST A MINUTE AND A HALF TO OPEN, THE CREW OF APOLLO 1 WAS DOOMED.

I'D BETTER NOT DESCRIBE WHAT I SEE.

NO ONE, ESPECIALLY NOT THE ASTRONAUTS THEMSELVES, HAD ANY ILLUSIONS ABOUT THE POTENTIAL DANGERS OF SPACEFLIGHT.

BOOOOM

GUS GRISSOM HAD ACKNOWLEDGED AS MUCH IN A PRESS CONFERENCE BEFORE THE ACCIDENT.

IF WE DIE, WE WANT PEOPLE TO ACCEPT IT. WE ARE IN A RISKY BUSINESS.

THE CONQUEST OF SPACE IS WORTH THE RISK OF LIFE.

BUT GRISSOM WAS TALKING ABOUT THE RISKS UP IN SPACE, NOT ON A LAUNCHPAD IN FLORIDA.

THE APOLLO 1 FIRE WAS SUCH A SHOCK BECAUSE THE TRAGEDY WASN'T CAUSED BY SOME EXOTIC HAZARD, BUT BY A SPARK DURING A ROUTINE TEST.

NASA GROUNDED ITS CREWS FOR MORE THAN A YEAR AND A HALF WHILE EVERYONE SCRUTINIZED APOLLO SPACECRAFT DESIGNS, ROOTING OUT ANY MORE POTENTIAL OVERSIGHTS.

WITH CONGRESS INVESTIGATING THE FIRE

AND AMERICAN TAXPAYERS GROWING SKITTISH AT THE PRICE TAG FOR THE SPACE PROGRAM,

NASA COULDN'T AFFORD ANOTHER TRAGEDY.

AFTER BREAKFAST, THE CREW BEGAN THE LONG, SLOW PROCESS OF DRESSING FOR SPACE.

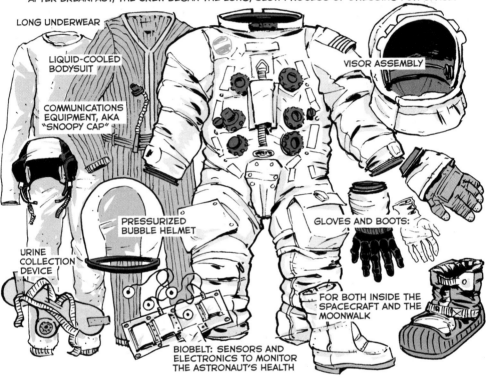

LONG UNDERWEAR

LIQUID-COOLED BODYSUIT

COMMUNICATIONS EQUIPMENT, AKA "SNOOPY CAP"

VISOR ASSEMBLY

URINE COLLECTION DEVICE

PRESSURIZED BUBBLE HELMET

GLOVES AND BOOTS:

FOR BOTH INSIDE THE SPACECRAFT AND THE MOONWALK

BIOBELT: SENSORS AND ELECTRONICS TO MONITOR THE ASTRONAUT'S HEALTH

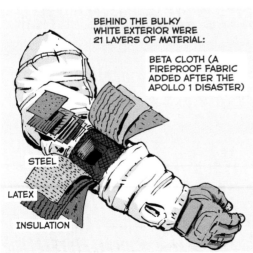

BEHIND THE BULKY WHITE EXTERIOR WERE 21 LAYERS OF MATERIAL:

BETA CLOTH (A FIREPROOF FABRIC ADDED AFTER THE APOLLO 1 DISASTER)

STEEL

LATEX

INSULATION

WITH ALL THE ACCESSORIES, THE APOLLO SUIT BECAME A PRESSURIZED PERSONAL SPACECRAFT.

A SPACECRAFT BUILT NOT BY AEROSPACE ENGINEERS, BUT BY A SUBSIDIARY OF THE BRA MANUFACTURER PLAYTEX.

THE SEAMSTRESSES AT THE INTERNATIONAL LATEX CORPORATION--THERE WERE NO MEN ON THE SEWING FLOOR AT ILC--ADAPTED THEIR STITCHING TO MEET NASA'S EXTREME REQUIREMENTS:

NO STITCH COULD STRAY MORE THAN 1/64TH OF AN INCH FROM THE SEAM,

AND FOR SEAMS THAT STRETCHED LONGER THAN A FOOTBALL FIELD, ANY OFF STITCH MEANT THAT THE WHOLE SUIT WOULD HAVE TO BE SCRAPPED.

WHILE MOLDMAKERS POURED CASTS OF THE ASTRONAUTS' BODIES TO GET A PERFECT FIT,

WORKERS WITH GLUE POTS LAMINATED HAIR-THIN SHEETS OF MYLAR FOIL.

EACH SPACE SUIT WAS LITERALLY HANDCRAFTED FROM 4000 COMPONENTS.

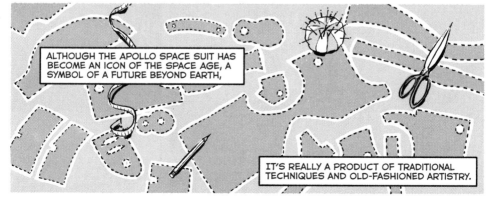

ALTHOUGH THE APOLLO SPACE SUIT HAS BECOME AN ICON OF THE SPACE AGE, A SYMBOL OF A FUTURE BEYOND EARTH,

IT'S REALLY A PRODUCT OF TRADITIONAL TECHNIQUES AND OLD-FASHIONED ARTISTRY.

EARLY PROTOTYPES FOR THE APOLLO SPACE SUIT TENDED TO BE BULKY, ARMORED SHELLS THAT LOOKED AS IF THEY WERE BORROWED FROM THE PAGES OF SCIENCE-FICTION COMIC BOOKS.

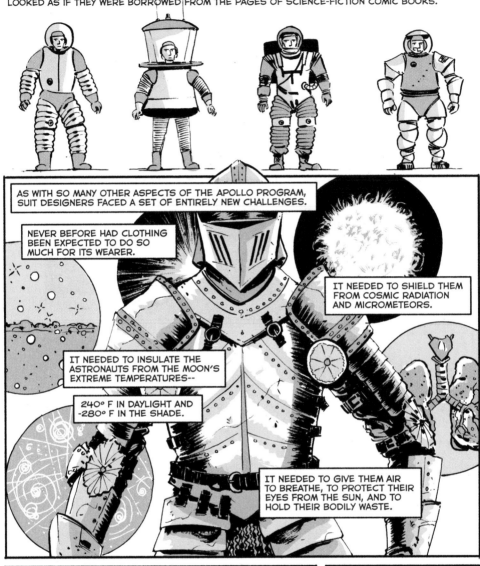

AS WITH SO MANY OTHER ASPECTS OF THE APOLLO PROGRAM, SUIT DESIGNERS FACED A SET OF ENTIRELY NEW CHALLENGES.

NEVER BEFORE HAD CLOTHING BEEN EXPECTED TO DO SO MUCH FOR ITS WEARER.

IT NEEDED TO SHIELD THEM FROM COSMIC RADIATION AND MICROMETEORS.

IT NEEDED TO INSULATE THE ASTRONAUTS FROM THE MOON'S EXTREME TEMPERATURES--

240° F IN DAYLIGHT AND -280° F IN THE SHADE.

IT NEEDED TO GIVE THEM AIR TO BREATHE, TO PROTECT THEIR EYES FROM THE SUN, AND TO HOLD THEIR BODILY WASTE.

AT THE SAME TIME, THE SUIT NEEDED TO BE FLEXIBLE SO THAT THE ASTRONAUTS COULD MOVE FREELY,

BOTH IN THE WEIGHTLESSNESS OF SPACE

AND ON THE ROCKY, DIFFICULT TERRAIN OF THE MOON.

THE FINAL DESIGN WEIGHED 183 POUNDS (WITH THE LIFE-SUPPORT BACKPACK),
THOUGH IN LUNAR GRAVITY IT WOULD FEEL MORE LIKE 30 POUNDS.

I'M GOING TO ATTEMPT TO COLLECT A SAMPLE--

%@#$

ARMSTRONG AND ALDRIN WOULD HAVE ONLY 160 MINUTES ON THE LUNAR SURFACE,

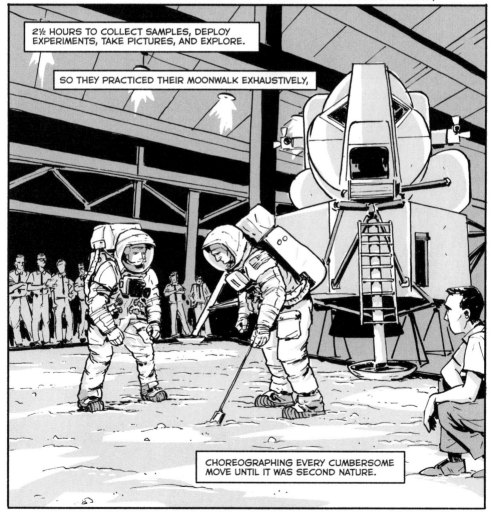

2½ HOURS TO COLLECT SAMPLES, DEPLOY EXPERIMENTS, TAKE PICTURES, AND EXPLORE.

SO THEY PRACTICED THEIR MOONWALK EXHAUSTIVELY,

CHOREOGRAPHING EVERY CUMBERSOME MOVE UNTIL IT WAS SECOND NATURE.

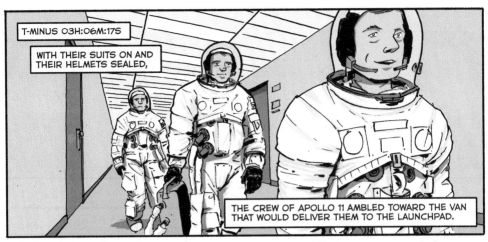

T-MINUS 03H:06M:17S

WITH THEIR SUITS ON AND THEIR HELMETS SEALED,

THE CREW OF APOLLO 11 AMBLED TOWARD THE VAN THAT WOULD DELIVER THEM TO THE LAUNCHPAD.

FIRE UP THOSE CAMERAS, BOYS, HERE THEY COME.

BY NOW THE ASTRONAUTS WERE USED TO THIS KIND OF CONSTANT SHUTTLING.

EVER SINCE SLAYTON HAD ASSIGNED THEM TO APOLLO 11, ARMSTRONG, ALDRIN, AND COLLINS WERE IN PERPETUAL MOTION,

EVERY MINUTE OF THEIR LIVES SCHEDULED AND TRACKED.

THERE WAS NEVER ENOUGH TIME.

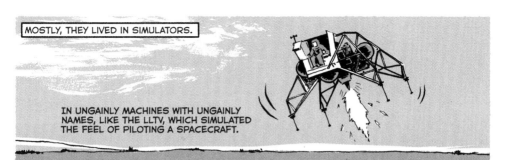

MOSTLY, THEY LIVED IN SIMULATORS.

IN UNGAINLY MACHINES WITH UNGAINLY NAMES, LIKE THE LLTV, WHICH SIMULATED THE FEEL OF PILOTING A SPACECRAFT.

OR THE POGO, WHICH SIMULATED LUNAR GRAVITY.

OR THE VOMIT COMET--AN AIRPLANE THAT INDUCED BRIEF MOMENTS OF WEIGHTLESSNESS.

THERE WERE SIMULATIONS TO HELP THE ASTRONAUTS GET THEIR BEARINGS IN SPACE,

SIMULATIONS TO PREPARE THEM FOR WHAT THEY MIGHT ENCOUNTER ON THE MOON,

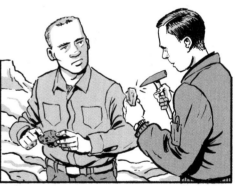

AND SIMULATIONS OF THEIR FIRST MOMENTS BACK ON EARTH.

FOR ARMSTRONG AND ALDRIN, ONE MACHINE IN PARTICULAR BECAME THEIR HOME AWAY FROM HOME.

THE LUNAR MODULE SIMULATOR,

AS CLOSE AS ANYONE COULD GET TO REPLICATING THE EXPERIENCE OF LANDING ON THE MOON.

INSTEAD OF WINDOWS, THE SIMULATOR HAD TV SCREENS SHOWING LIVE FOOTAGE FROM THE NEXT ROOM OVER,

WHERE A CAMERA FILMED THE MOLDED CRATERS OF A RUBBER MOON MODEL.

FOR ADDED REALISM, THE CREW WAS IN RADIO CONTACT WITH COLLINS IN HIS OWN SIMULATOR AND WITH GENE KRANZ AND HIS CONTROLLERS IN HOUSTON.

APOLLO 11 IS *GO* FOR SEPARATION.

ROGER. MIKE, WHAT'S YOUR PITCH ANGLE?

004 DEGREES.

THE MEN WHO RAN THE SIMULATIONS DID THEIR BEST TO MAKE LANDING ON THE MOON AS HARD AS POSSIBLE.

ALL DAY LONG THEY CONJURED UP EVERY CONTINGENCY THAT THE ASTRONAUTS MIGHT ENCOUNTER.

OK, LET'S CUT POWER TO THRUSTERS 5 AND 9.

IN THE SIMULATOR, MOST MISSIONS ENDED BADLY.

ABORT! *ABORT!*

KNOCK, KNOCK.

BY MID-JUNE, A MONTH BEFORE APOLLO 11'S LAUNCH DATE, THIS RELENTLESS PACE WAS STARTING TO TAKE A TOLL.

WASHINGTON IS TALKING ABOUT MOVING BACK THE LAUNCH.

THEY WANTED ME TO SEE IF YOU GUYS ARE READY.

THE CREW UNDERSTOOD THE SUBTEXT:

A DELAY AT THIS POINT WOULDN'T MEAN WAITING AN EXTRA DAY OR TWO, IT WOULD MEAN AN ENTIRE MONTH.

THAT'S BECAUSE OF HOW NASA CALCULATED THE LAUNCH WINDOW-- THE IDEAL TIME TO SEND ASTRONAUTS TO THE MOON.

TO START WITH THE OBVIOUS:

THE ASTRONAUTS WOULD NEED TO SEE WHAT THEY WERE DOING,

SO THEY OUGHT TO LAND ON THE MOON WHEN THE SUN IS UP.

THE LUNAR MORNING IS BEST, WHEN THE LOW ANGLE OF THE LIGHT MAKES CRATERS AND BOULDERS EASY TO SPOT.

FOR ANY GIVEN LONGITUDE ON THE MOON, SUCH CONDITIONS HAPPEN ONLY ONCE A MONTH.

THEN THERE IS THE QUESTION OF *WHERE* TO SEND THE ASTRONAUTS.

FIRST, THE LANDING SITE SHOULD BE FLAT AND FREE OF HAZARDS.

SECOND, IF IT'S AT THE RIGHT LATITUDE, IT WON'T REQUIRE MUCH EXTRA MANEUVERING-- I.E., PROPELLANT--TO REACH.

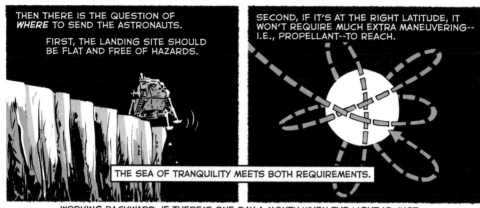

THE SEA OF TRANQUILITY MEETS BOTH REQUIREMENTS.

WORKING BACKWARD, IF THERE'S ONE DAY A MONTH WHEN THE LIGHT IS JUST RIGHT, AND IF IT TAKES THREE DAYS TO GET FROM THE EARTH TO THE MOON,

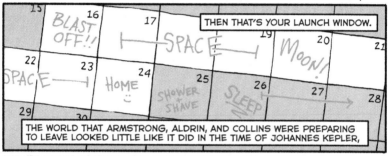

THEN THAT'S YOUR LAUNCH WINDOW.

THE WORLD THAT ARMSTRONG, ALDRIN, AND COLLINS WERE PREPARING TO LEAVE LOOKED LITTLE LIKE IT DID IN THE TIME OF JOHANNES KEPLER,

AND YET, IN ONE ESSENTIAL WAY, THREE CENTURIES HAD CHANGED NOTHING:

TO THE MOON, AND ITS AGE-OLD ARC ACROSS THE HEAVENS, THESE MEN WERE INESCAPABLY BOUND.

SO, GUYS, HOW DO YOU FEEL?

I'M READY, DEKE.

WE COULD TRAIN FOR ANOTHER YEAR AND STILL NOT FIX EVERY LITTLE PROBLEM.

WE SPEND TOO MUCH TIME DOING THAT AND WE'LL FORGET WHAT THIS MISSION IS ALL ABOUT.

IT'LL BE CLOSE.

BUT WE'LL MAKE IT.

TELL HEADQUARTERS NOT TO WORRY.

WHILE HIS CREWMATES TOOK THEIR SEATS, ALDRIN LINGERED AT THE LAUNCH TOWER RAILING.

DAWN WAS BREAKING OVER THE ATLANTIC OCEAN.

"I WAITED ALONE," HE WOULD LATER RECALL, "IN A SORT OF SERENE LIMBO."

ABOVE HIM STRETCHED INFINITY,

AND BELOW HIS FEET:

"THAT WONDROUS WHITE MACHINE THAT WAS GOING TO PROPEL US OFF INTO HISTORY--WE HOPED."

AT THE THRESHOLD OF THE SPACECRAFT STOOD PAD LEADER GÜNTER WENDT, A JOVIAL ENGINEER WHO'D BEEN LOADING ASTRONAUTS ONTO ROCKETS SINCE PROJECT MERCURY.

MR. ARMSTRONG, WE THOUGHT YOU MIGHT NEED THIS...

IT'S A KEY TO THE MOON!

FREE RIDE
"SPACE TAXI"
GOOD FOR ONE ROUND TRIP FREE

HA! HERE, GÜNTER, I'LL TRADE YOU.

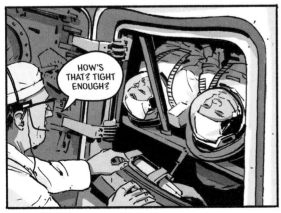

HOW'S THAT? TIGHT ENOUGH?

GODSPEED, APOLLO 11.

T-MINUS 00H:55M:42S

MILES AWAY, NEARLY A MILLION PEOPLE HAD GATHERED ON THE BEACHES AND ALONG THE HIGHWAYS.

THEY CAME TO WITNESS, TO SAY THAT THEY HAD SEEN THIS MOMENT WITH THEIR OWN EYES.

NASA'S OWN GUEST LIST WAS 20,000 NAMES LONG.

REPORTERS BROADCAST THE SPECTACLE TO AT LEAST A SIXTH OF THE EARTH'S POPULATION.

IT'S A TIME OF EXHILARATION, REFLECTION, HOPE, FULFILLMENT, AS A CENTURIES-OLD DREAM STARTS TOWARD REALITY.

JACK KING, THE NASA PUBLIC AFFAIRS OFFICER, COUNTED DOWN THE MINUTES TO LAUNCH.

ALL SYSTEMS CONTINUING TO LOOK GOOD AT THIS POINT.

WE ARE STILL AIMING TOWARD OUR PLANNED LIFTOFF AT THE START OF THE LUNAR WINDOW, 9:32 A.M.

FOR NOW, EVERYONE WAITED.

T-MINUS OOH:03M:57S

ALL RIGHT-- CMP ON PANEL 3 INSERT VERB 75...

VERB 75 STANDING BY.

THE ONLY WORDS WE HAVE,

START SEQUENCE INITIATED.

THAT COME CLOSE TO DESCRIBING THE FURY

PAD COMM GOING OFF.

COLLINS

AND THE FIRE OF WHAT WAS ABOUT TO COME,

SURE HAS BEEN A NICE, SMOOTH COUNTDOWN.

ARMSTRONG

AW, THANK YOU, BABE.

IGNITION SEQUENCE START...

...THREE....

...TWO...

...ONE...

ARE WORDS THAT BELONG IN TALES OF MYTHIC AND IMPOSSIBLE THINGS.

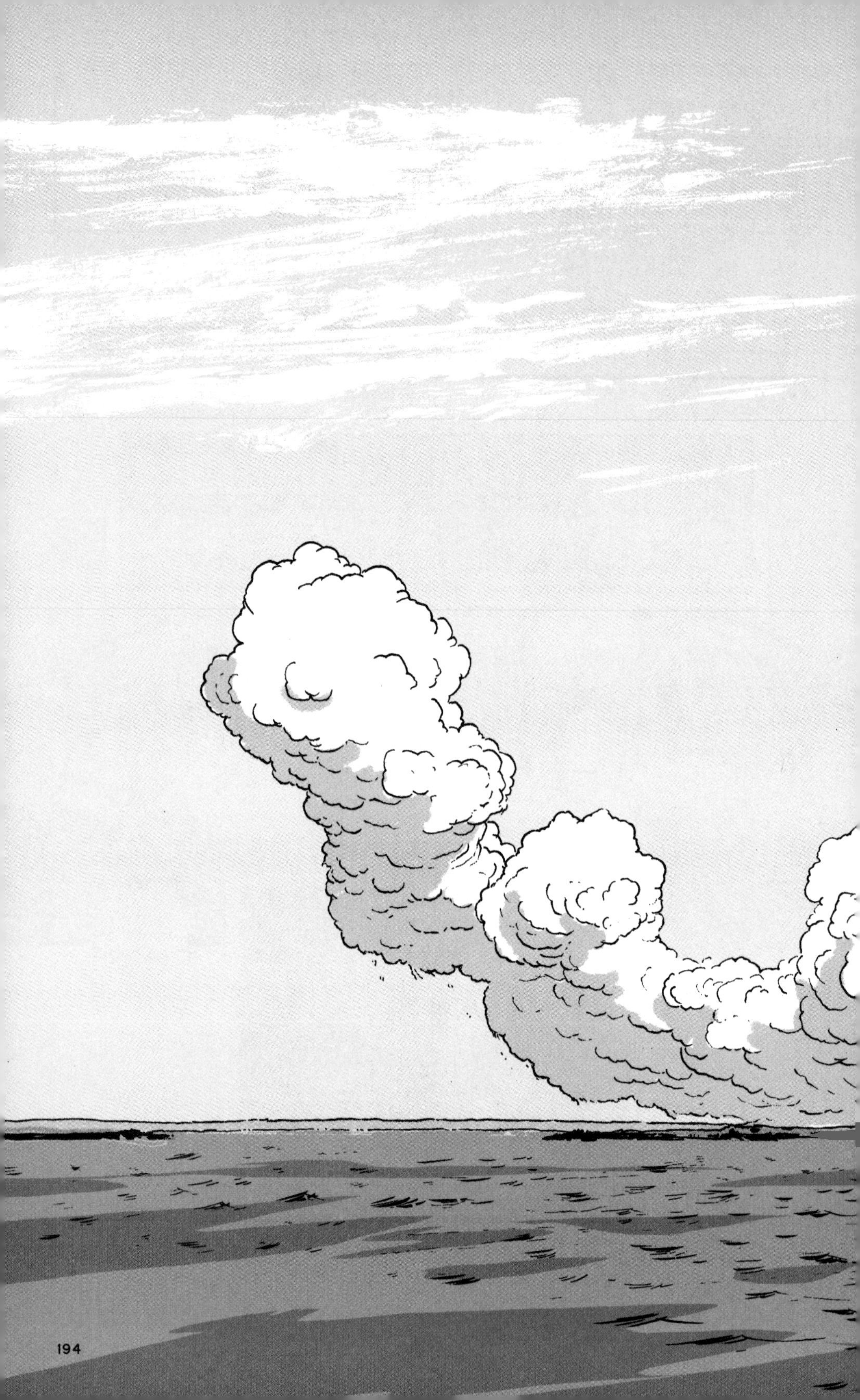

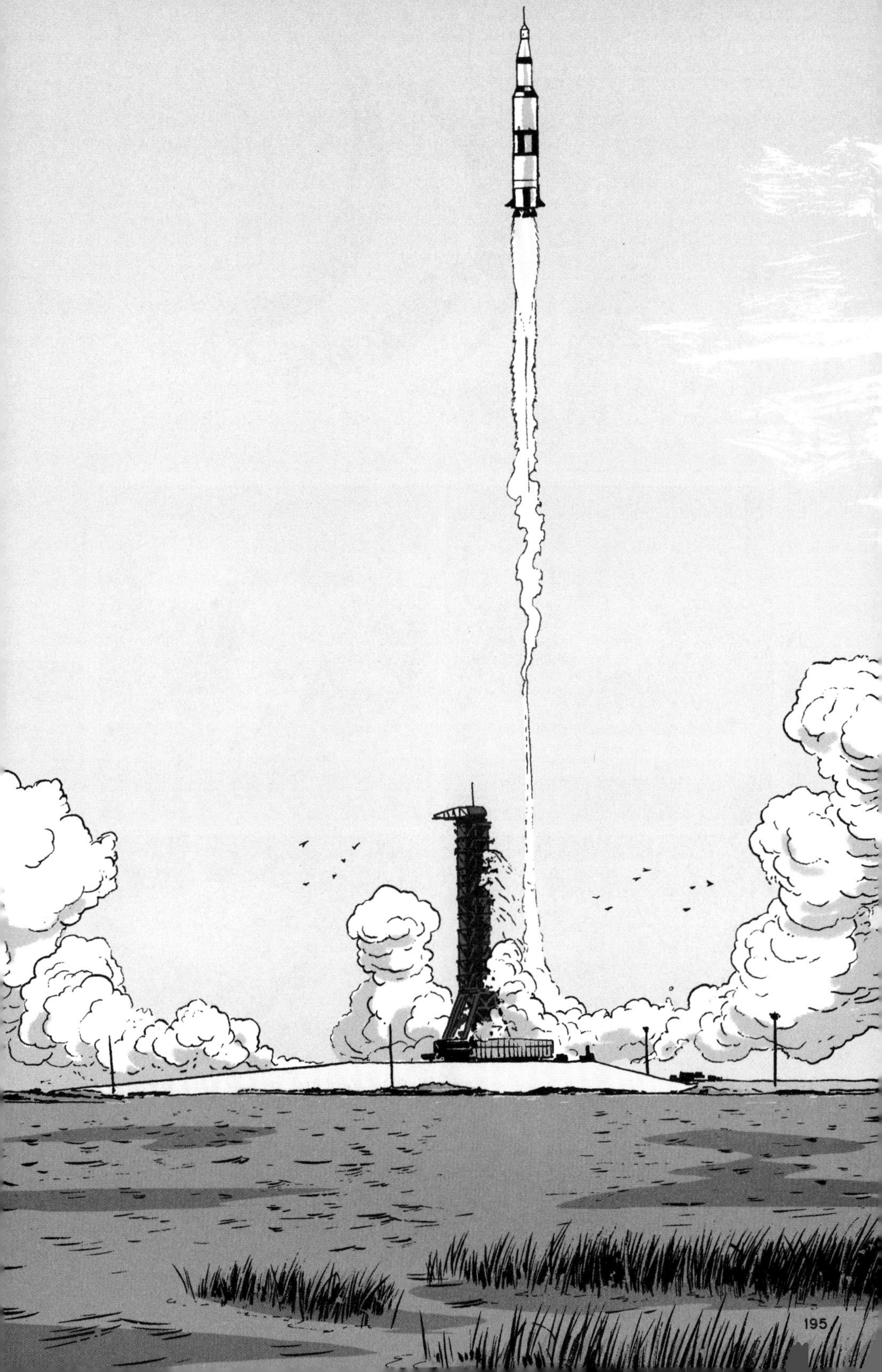

IN TWO MINUTES THE FIRST STAGE OF THE SATURN V BURNED THROUGH ALMOST FIVE MILLION POUNDS OF PROPELLANT.

APOLLO 11, HOUSTON. YOU'RE GO FOR STAGING.

INBOARD CUTOFF.

AND THEN, 40 MILES ABOVE THE EARTH, ITS JOB WAS DONE.

STAGING.

RATHER THAN CARRY DEADWEIGHT,

ARMSTRONG KEYED IN THE COMMAND TO DETONATE A RING OF EXPLOSIVES,

SHEARING OFF THE EMPTY HULL OF THE FIRST STAGE, LEAVING IT TO TUMBLE BACK DOWN INTO THE ATLANTIC OCEAN.

APOLLO 11 KEPT CLIMBING, LIGHTER NOW, GAINING THE BETTER PART OF A MILE EVERY SECOND.

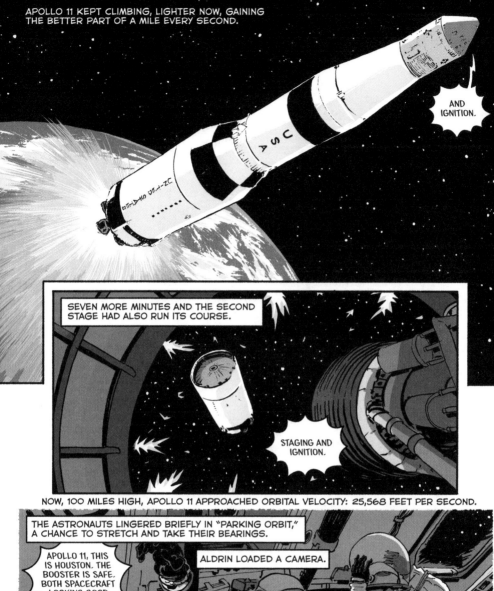

AND IGNITION.

SEVEN MORE MINUTES AND THE SECOND STAGE HAD ALSO RUN ITS COURSE.

STAGING AND IGNITION.

NOW, 100 MILES HIGH, APOLLO 11 APPROACHED ORBITAL VELOCITY: 25,568 FEET PER SECOND.

THE ASTRONAUTS LINGERED BRIEFLY IN "PARKING ORBIT," A CHANCE TO STRETCH AND TAKE THEIR BEARINGS.

APOLLO 11, THIS IS HOUSTON. THE BOOSTER IS SAFE. BOTH SPACECRAFT LOOKING GOOD.

ALDRIN LOADED A CAMERA.

GET A PICTURE OF THAT!

GOD DAMN, THAT'S PRETTY! THIS IS UNREAL. I'D FORGOTTEN.

THEY'D ALL SEEN IT BEFORE, THE EARTH FROM THIS HEAVENLY PERCH, BUT THE VIEW WAS NEVER NOT PROFOUND.

ONE OF NASA'S LESS OBVIOUS INNOVATIONS WAS THE NEW LANGUAGE ITS ENGINEERS AND ASTRONAUTS HAD INVENTED, A JARGON OF ACRONYMS AND ABBREVIATIONS.

YOU SHOULD STABILIZE AND ALIGN CM-BMAG MODE 3 TO ATT 1/ RATE 2

NASA-SPEAK WAS QUICK AND EFFICIENT, PERFECT FOR THE TECHNICAL ASPECTS OF SPACEFLIGHT.

BUT AS THE ROLE OF ASTRONAUT EVOLVED FROM TEST PILOT TO EXPLORER, THEIR LANGUAGE TOOK ON AN ADDED BURDEN.

FOR ALL OF US WHO WERE AND WOULD ALWAYS BE EARTHBOUND, THE ASTRONAUTS BECAME OUR TOUR GUIDES TO THE WONDERS OF SPACE.

IN THIS THEY WERE DISAPPOINTINGLY TONGUE-TIED.

A BEAUTIFUL SIGHT!

IT'S BEEN A BEAUTIFUL VIEW!

WHAT A VIEW, BY GOLLY!

SURE IS BEAUTIFUL OUT HERE!

WHEN PRESSED, THE ASTRONAUTS WERE UNAPOLOGETIC.

OF COURSE I THOUGHT ABOUT THE MAGNIFICENCE OF THE WHOLE THING,

BUT THAT'S DIFFICULT TO CAPTURE IN A SIMPLE DESCRIPTION.

WE WEREN'T TRAINED TO EMOTE, WE WERE TRAINED TO REPRESS EMOTIONS

LEST THEY INTERFERE WITH OUR VERY COMPLICATED, DELICATE, AND ONE-CHANCE-ONLY DUTIES.

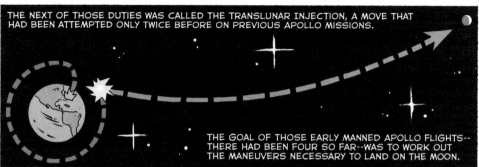

THE NEXT OF THOSE DUTIES WAS CALLED THE TRANSLUNAR INJECTION, A MOVE THAT HAD BEEN ATTEMPTED ONLY TWICE BEFORE ON PREVIOUS APOLLO MISSIONS.

THE GOAL OF THOSE EARLY MANNED APOLLO FLIGHTS-- THERE HAD BEEN FOUR SO FAR--WAS TO WORK OUT THE MANEUVERS NECESSARY TO LAND ON THE MOON.

APOLLO 7:

A TRIAL RUN IN EARTH ORBIT OF THE COMMAND AND SERVICE MODULE,

THE ASTRONAUTS' HOME AWAY FROM HOME FOR THE MOONSHOT.

WOW, BOY. WE OUGHTA QUIT WHILE WE'RE AHEAD.

APOLLO 8:

THE FIRST FORAY INTO LUNAR ORBIT, A CHANCE TO PRACTICE NAVIGATION AND TO SCOUT LANDING SITES.

AND GOD SAW THE LIGHT, THAT IT WAS GOOD: AND GOD DIVIDED THE LIGHT FROM THE DARKNESS.

APOLLO 9:

BACK IN EARTH ORBIT, THE FIRST FLIGHT OF THE LUNAR MODULE.

BUT YOU ARE UPSIDE DOWN AGAIN!

I WAS JUST THINKING ONE OF US ISN'T RIGHT SIDE UP.

APOLLO 10:

A SHAKEDOWN CRUISE, THE DRESS REHEARSAL FOR THE MOON LANDING.

WE IS DOWN AMONG THEM, CHARLIE!

BY THE TIME IT WAS APOLLO 11'S TURN, NEARLY EVERY ASPECT OF THE MISSION HAD BEEN TESTED.

THE ONLY REMAINING UNKNOWN WAS ALSO THE RISKIEST: THE LANDING ITSELF.

199

THREE HOURS INTO THEIR FLIGHT, COASTING ON A TRAJECTORY THAT WOULD SWING THEM AROUND THE FAR SIDE OF THE MOON,

APOLLO 11, THIS IS HOUSTON.

YOU ARE GO FOR SEPARATION.

THE ASTRONAUTS WERE READY TO JETTISON THE LAST STAGE OF THE SATURN V.

THE ROCKET MAY HAVE RUN ITS COURSE, BUT IT STILL HELD PRECIOUS CARGO:

THE LUNAR MODULE, CALL SIGN EAGLE

THE FIRST TRUE SPACESHIP.

BECAUSE THE MOON HAS NO ATMOSPHERE, THERE WAS NO NEED FOR THE AERODYNAMICS OF AN AIRPLANE.

THE LM WAS DESIGNED WITH A FORM TO FOLLOW ITS OTHERWORLDLY FUNCTION:

DELICATE ANTENNAS,

HAIR-THIN WALLS,

A SKIRT OF INSULATING FOIL,

SHOCK-ABSORBING LEGS,

AND, MOST IMPORTANT FOR THE TASK AHEAD, A LADDER.

COLUMBIA COULD CONNECT TO THE EAGLE THROUGH A HATCH ON THE ROOF.

STAND BY. WE'RE CLOSING.

BUT FIRST THEY HAD TO DOCK.

WITH THE TWO MODULES JOINED TOGETHER AND THE DISCARDED REMNANTS OF THE SATURN ROCKET ON A COLLISION COURSE WITH THE SUN,

THE CREW OF APOLLO 11 SETTLED IN FOR THE THREE-DAY VOYAGE INTO LUNAR ORBIT.

CECIL B. DE ALDRIN IS STANDING BY FOR INSTRUCTIONS.

ALONG THE WAY, NASA HAD THEM FIRE UP THE ONBOARD TV CAMERA,

TO GIVE THE AUDIENCE BACK HOME A PEEK AT THE TOPSY-TURVY WORLD OF ZERO G.

WE GOT THE NETWORK ALL CONFIGURED FOR THE TV. YOU CAN START ANYTIME YOU WANT. OVER.

THERE'S PLENTY OF ROOM UP HERE FOR THE THREE OF US, BUT AFTER A WHILE YOU GET TIRED OF RATTLING AROUND, BANGING OFF THE CEILING AND FLOOR, SO YOU FIND A LITTLE CORNER SOMEWHERE...

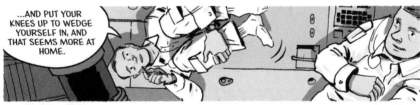

...AND PUT YOUR KNEES UP TO WEDGE YOURSELF IN, AND THAT SEEMS MORE AT HOME.

WELL, I GUESS IT'S ALMOST YOUR DINNERTIME DOWN THERE, EARTH.

WE'VE GOT ALL KINDS OF GOOD STUFF. COFFEE, BACON...

WOULD YOU BELIEVE YOU'RE LOOKING AT CHICKEN STEW HERE?

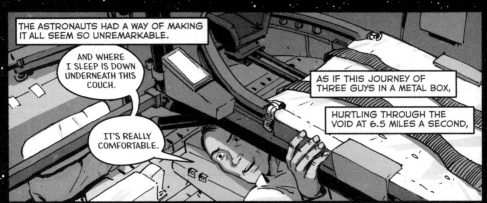

THE ASTRONAUTS HAD A WAY OF MAKING IT ALL SEEM SO UNREMARKABLE.

AND WHERE I SLEEP IS DOWN UNDERNEATH THIS COUCH.

IT'S REALLY COMFORTABLE.

AS IF THIS JOURNEY OF THREE GUYS IN A METAL BOX,

HURTLING THROUGH THE VOID AT 6.5 MILES A SECOND,

WERE JUST LIKE ANY OLD ROAD TRIP.

THAT ILLUSION DISSOLVED, THOUGH, WHENEVER ANYONE LOOKED OUT THE WINDOW,

AT THE EARTH, SHRINKING BY THE HOUR.

SMALL AND BRIGHT AND BLUE AND SINGULAR.

FROM 100,000 MILES AWAY THE EARTH LOOKED TRANQUIL.

BUT OF COURSE THAT WAS
JUST ANOTHER ILLUSION.

THERE WAS NOTHING TRANQUIL ABOUT THE
WORLD THAT APOLLO 11 HAD LEFT BEHIND.

OR MAYBE IT WASN'T AN ILLUSION,

BUT A KIND OF EPIPHANY:

HERE WAS THE SUM TOTAL
OF HUMANKIND, ALL THAT
HOME HAS EVER MEANT,

SHRUNK BY DISTANCE TO
THE SIZE OF A THUMBPRINT.

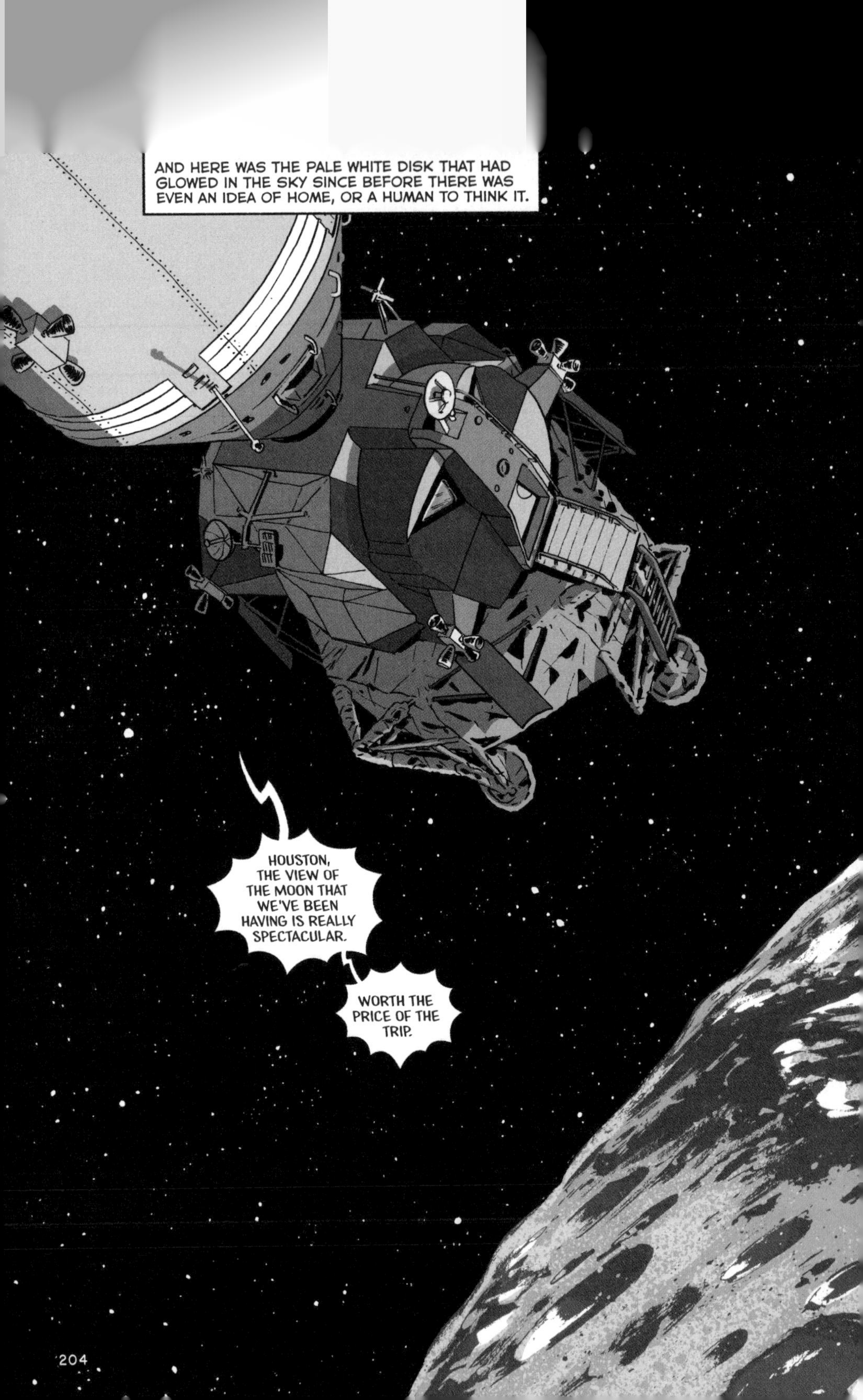
AND HERE WAS THE PALE WHITE DISK THAT HAD GLOWED IN THE SKY SINCE BEFORE THERE WAS EVEN AN IDEA OF HOME, OR A HUMAN TO THINK IT.
HOUSTON, THE VIEW OF THE MOON THAT WE'VE BEEN HAVING IS REALLY SPECTACULAR.
WORTH THE PRICE OF THE TRIP.

UP CLOSE, RAKED BY EARTHSHINE AND SUNLIGHT,

THE MOON BULGED IN A WAY THAT
CAUGHT THE ASTRONAUTS OFF GUARD.

LIKE THE REST OF US, THEY HAD ONLY
EVER SEEN THE MOON IN TWO DIMENSIONS,

ITS DEPTH LOST IN TRANSLATION TO THE
PRINTED PAGE OR ERASED BY DISTANCE.

COLLINS THOUGHT: I ALMOST FEEL
I CAN REACH OUT AND TOUCH IT.

SOON, THE MEN FLOATING BESIDE
HIM WOULD DO JUST THAT.

CHAPTER XVII
TRANQUILITY

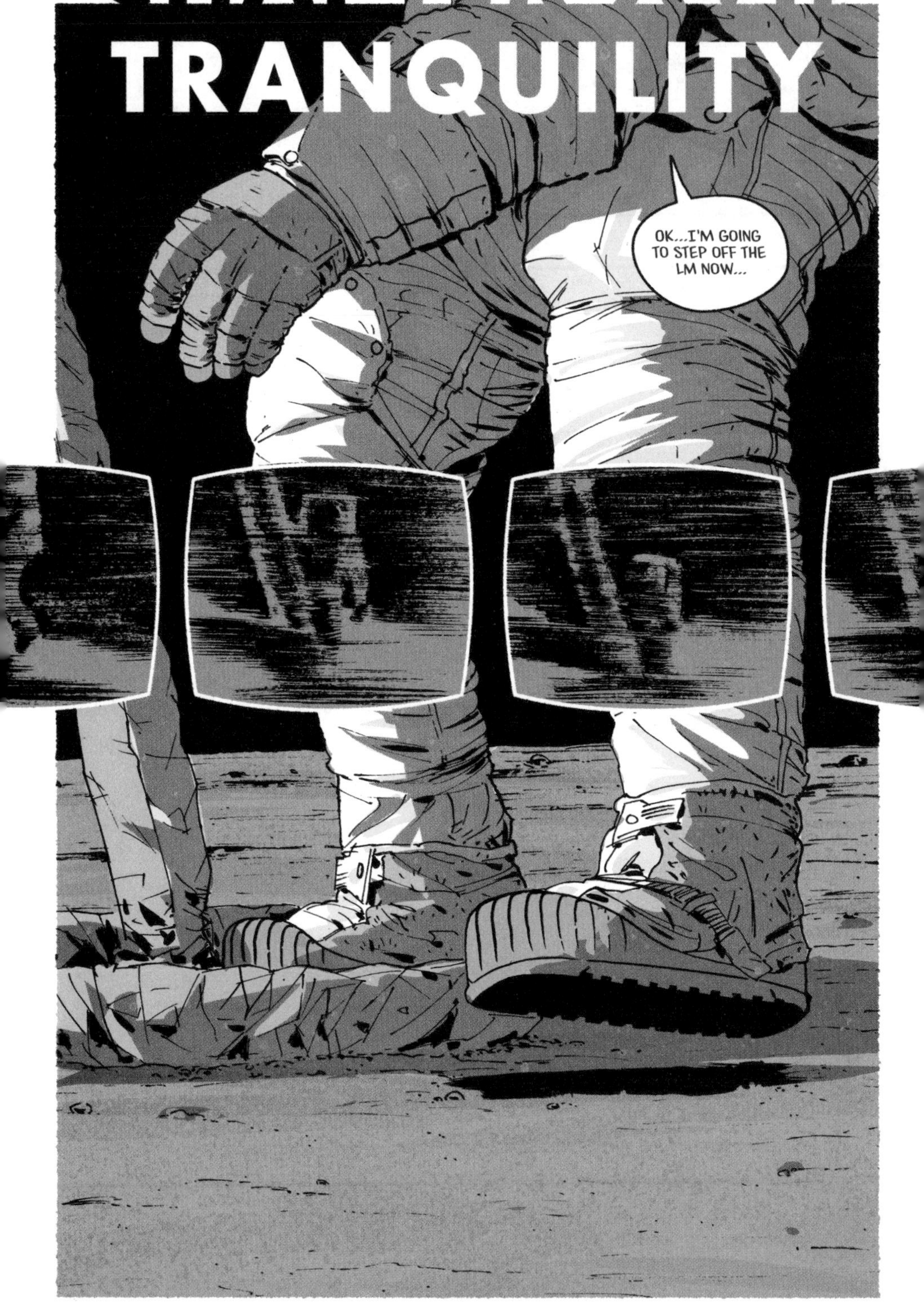
OK...I'M GOING TO STEP OFF THE LM NOW...

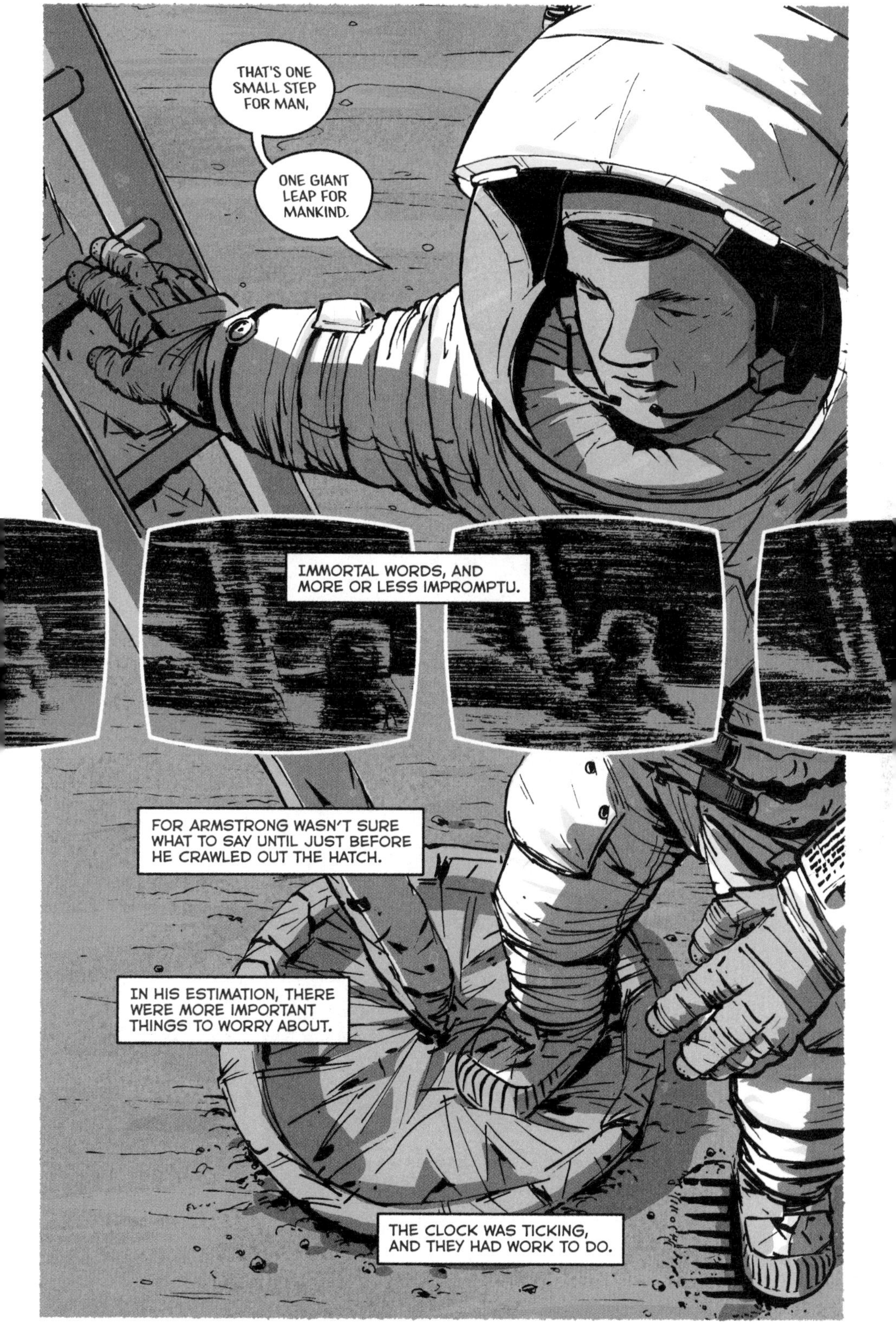
THAT'S ONE SMALL STEP FOR MAN,
ONE GIANT LEAP FOR MANKIND.
IMMORTAL WORDS, AND MORE OR LESS IMPROMPTU.
FOR ARMSTRONG WASN'T SURE WHAT TO SAY UNTIL JUST BEFORE HE CRAWLED OUT THE HATCH.
IN HIS ESTIMATION, THERE WERE MORE IMPORTANT THINGS TO WORRY ABOUT.
THE CLOCK WAS TICKING, AND THEY HAD WORK TO DO.

AT THIS POINT, IF SOMETHING WENT HAYWIRE,

THE SURFACE IS FINE AND POWDERY.

I CAN KICK IT UP LOOSELY WITH MY FOOT.

THERE SEEMS TO BE NO DIFFICULTY IN MOVING AROUND.

IT'S EVEN PERHAPS EASIER THAN THE SIMULATIONS.

AND THE ASTRONAUTS HAD TO LEAVE THE MOON IN A HURRY,

OK. GOING TO GET THE CONTINGENCY SAMPLE THERE, NEIL?

RIGHT.

THEY WOULD AT LEAST HAVE THIS SMALL SAMPLE OF LUNAR SOIL TO BRING HOME.

AT FIRST THERE WAS SOME DEBATE OVER WHO WOULD GET TO MAKE THE FIRST STEP ONTO THE MOON.

ARE YOU READY FOR ME TO COME OUT?

ALL SET.

ULTIMATELY, WHAT IT CAME DOWN TO WAS A HINGE.

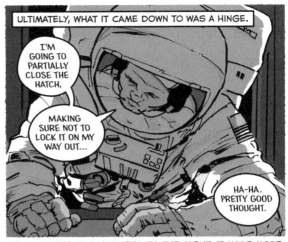

I'M GOING TO PARTIALLY CLOSE THE HATCH,

MAKING SURE NOT TO LOCK IT ON MY WAY OUT...

HA-HA. PRETTY GOOD THOUGHT.

BECAUSE THE HATCH HINGED TO THE RIGHT, IT MADE MORE SENSE FOR WHOEVER WAS ON THE LEFT TO EXIT FIRST.

SO ARMSTRONG IT WOULD BE.

YOU'VE GOT THREE MORE STEPS, THEN A LONG ONE.

BEAUTIFUL VIEW!

ISN'T THAT SOMETHING. MAGNIFICENT SIGHT OUT HERE.

MAGNIFICENT DESOLATION.

WHILE ARMSTRONG SET UP A CAMERA TRIPOD, ALDRIN UNLOADED THE LM'S CARGO.

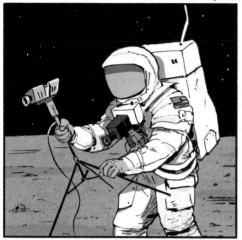

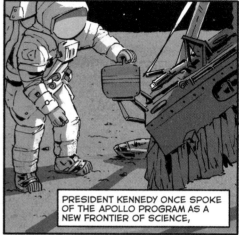

PRESIDENT KENNEDY ONCE SPOKE OF THE APOLLO PROGRAM AS A NEW FRONTIER OF SCIENCE,

AND YET THE CHALLENGE OF LAUNCHING ASTRONAUTS TO THE MOON TURNED OUT TO BE MOSTLY AN EXERCISE IN ENGINEERING.

BUT NOW AND FOR THE NEXT HOUR, NASA'S SCIENTISTS WOULD HAVE THEIR MOMENT.

ALDRIN REMOVED THREE CAREFULLY DESIGNED EXPERIMENTS FROM THE BELLY OF THE *EAGLE*:

A SHEET OF ALUMINUM FOIL TO CAPTURE CHARGED PARTICLES FROM THE SUN;

A SEISMOMETER TO MEASURE MOONQUAKES AND CHART THE INTERNAL STRUCTURE OF THE MOON;

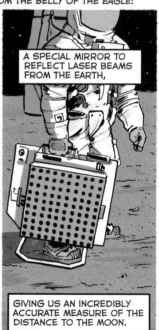

A SPECIAL MIRROR TO REFLECT LASER BEAMS FROM THE EARTH,

GIVING US AN INCREDIBLY ACCURATE MEASURE OF THE DISTANCE TO THE MOON.

EVEN SOME OF THE MOST ICONIC IMAGES FROM APOLLO 11 ARE ARTIFACTS OF THESE EXPERIMENTS.

THE SOIL IS VERY COHESIVE. IT WILL RETAIN A SLOPE OF PROBABLY 70 DEGREES ALONG THE SIDE...

HOPING TO AVOID THE IMPRESSION THAT AMERICANS WERE MAKING A COSMIC LAND-GRAB,

NASA INITIALLY PLANNED FOR THE ASTRONAUTS TO PLANT THE FLAG OF THE UNITED NATIONS.

TRANQUILITY BASE, THIS IS HOUSTON...

INSTEAD, APOLLO 11 CARRIED A $5 NYLON STARS AND STRIPES.

...COULD WE GET BOTH OF YOU ON THE CAMERA FOR A MINUTE, PLEASE?

HELLO, NEIL AND BUZZ.

I'M TALKING TO YOU BY TELEPHONE FROM THE OVAL ROOM AT THE WHITE HOUSE.

I JUST CAN'T TELL YOU HOW PROUD WE ALL ARE.

BECAUSE OF WHAT YOU HAVE DONE, THE HEAVENS HAVE BECOME A PART OF MAN'S WORLD.

HOUSTON, *COLUMBIA* ON THE HIGH GAIN ANTENNA. HOW'S IT GOING?

I GUESS YOU'RE ABOUT THE ONLY PERSON AROUND THAT DOESN'T HAVE TV COVERAGE OF THE SCENE.

THE EVA* IS PROGRESSING BEAUTIFULLY.

THAT'S ALL RIGHT. HOW'S THE QUALITY?

*EXTRAVEHICULAR ACTIVITY

IS THE LIGHTING HALFWAY DECENT?

YES, INDEED.

HOUSTON, HOW DOES OUR TIME LINE APPEAR TO BE GOING?

IT LOOKS LIKE YOU'RE ABOUT A HALF HOUR SLOW ON IT.

IT WAS TIME TO START WRAPPING THINGS UP. THERE WERE STILL CORE SAMPLES AND ROCKS TO COLLECT FOR THE GEOLOGISTS BACK HOME.

BUT FIRST, ARMSTRONG WANTED TO TAKE JUST A FEW MORE PHOTOGRAPHS.

AFTER TRAVELING 240,000 MILES TO GET TO THE MOON, IT MIGHT SEEM ODD THAT THE ASTRONAUTS STAYED WITHIN 50 YARDS OF THE LM.

BUT THEY WERE BEING CAUTIOUS.

THIS WAS, AFTER ALL, UNCHARTED TERRITORY.

DISTANCES ON THE MOON WERE HARD TO GAUGE.

THE HORIZON SEEMED TO DROP OFF TOO SOON,

AND PEBBLES SEEN UP CLOSE TURNED OUT TO BE BOULDERS.

FROM UP ON THE RIM OF A CRATER, ARMSTRONG GLANCED BACK AT HIS CREWMATE AND THEIR SHIP,

BACK AT THE SITE WHERE, FOR 22 HOURS, THE COLD STILLNESS OF THIS DEAD PLANET WAS BRIEFLY BROKEN.

213

ALMOST AS SOON AS IT STARTED, IT SEEMED, THEIR FORAY OUT ON THE SEA OF TRANQUILITY WAS OVER.

BUZZ, THIS IS HOUSTON.

IT'S ABOUT TIME FOR YOU TO START YOUR CLOSEOUT ACTIVITIES.

ROGER.

LATER, ALDRIN WOULD REFLECT:

"THERE WAS FAR MORE TO INVESTIGATE THAN WE COULD EVER HOPE TO COVER."

"WE DIDN'T EVEN SCRATCH THE SURFACE."

ADIOS, AMIGO.

ARMSTRONG AND ALDRIN HAD BEEN AWAKE FOR ALMOST A DAY.

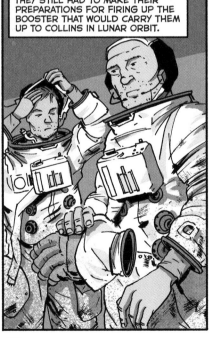

THEY STILL HAD TO MAKE THEIR PREPARATIONS FOR FIRING UP THE BOOSTER THAT WOULD CARRY THEM UP TO COLLINS IN LUNAR ORBIT.

BUT ALL THAT COULD COME LATER.

FIRST, THEY NEEDED SLEEP.

...TRANQUILITY BASE, TRANQUILITY BASE...

GOOD MORNING, HOUSTON.

GOT A COUPLE CHANGES TO YOUR SURFACE CHECKLIST HERE.

ROGER

LATER, AFTER TWO SLOW HOURS OF PROCEDURE:

THREE...TWO... ONE...

WE'RE OFF!

LOOK AT THAT STUFF GO ALL OVER THE PLACE.

AND LOOK AT THAT SHADOW.

BEAUTIFUL!

IF, EVENTUALLY, EXPLORERS
FROM FARTHER AWAY THAN EARTH

ARE DILIGENT ENOUGH TO MAKE
SENSE OF WHAT WAS LEFT BEHIND,

TO SORT THROUGH THE
SCRAPS OF PLASTIC,

THE DISCARDED CLOTHING
AND BAGS OF URINE,

THE BITS OF METAL AND
ABANDONED TOOLS,

THEY MIGHT LINGER ON THE
SMALL PLAQUE STRAPPED
TO THE LADDER RUNGS:

"HERE MEN FROM THE PLANET EARTH FIRST
SET FOOT UPON THE MOON. JULY 1969, A.D.

"WE CAME IN PEACE FOR ALL MANKIND."

CHAPTER XVIII
EARTHBOUND

THE CREW WAS ABOUT TO PERFORM A CARGO SWAP: GARBAGE AND ANYTHING UNNECESSARY FROM THE *COLUMBIA* IN EXCHANGE FOR THE *EAGLE'S* LUNAR SAMPLES.

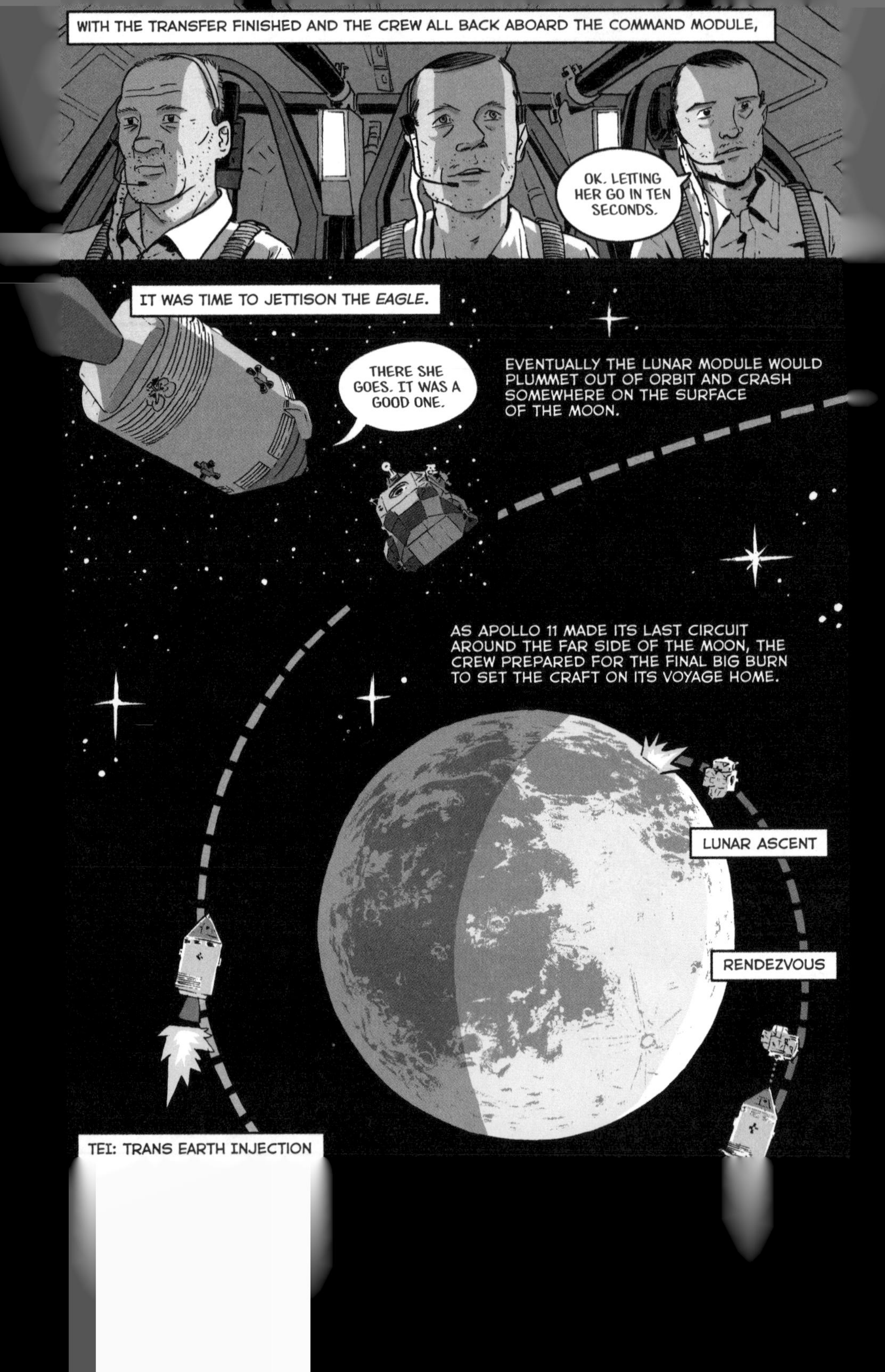
WITH THE TRANSFER FINISHED AND THE CREW ALL BACK ABOARD THE COMMAND MODULE,
OK. LETTING HER GO IN TEN SECONDS.
IT WAS TIME TO JETTISON THE *EAGLE*.
THERE SHE GOES. IT WAS A GOOD ONE.
EVENTUALLY THE LUNAR MODULE WOULD PLUMMET OUT OF ORBIT AND CRASH SOMEWHERE ON THE SURFACE OF THE MOON.
AS APOLLO 11 MADE ITS LAST CIRCUIT AROUND THE FAR SIDE OF THE MOON, THE CREW PREPARED FOR THE FINAL BIG BURN TO SET THE CRAFT ON ITS VOYAGE HOME.
LUNAR ASCENT
RENDEZVOUS
TEI: TRANS EARTH INJECTION

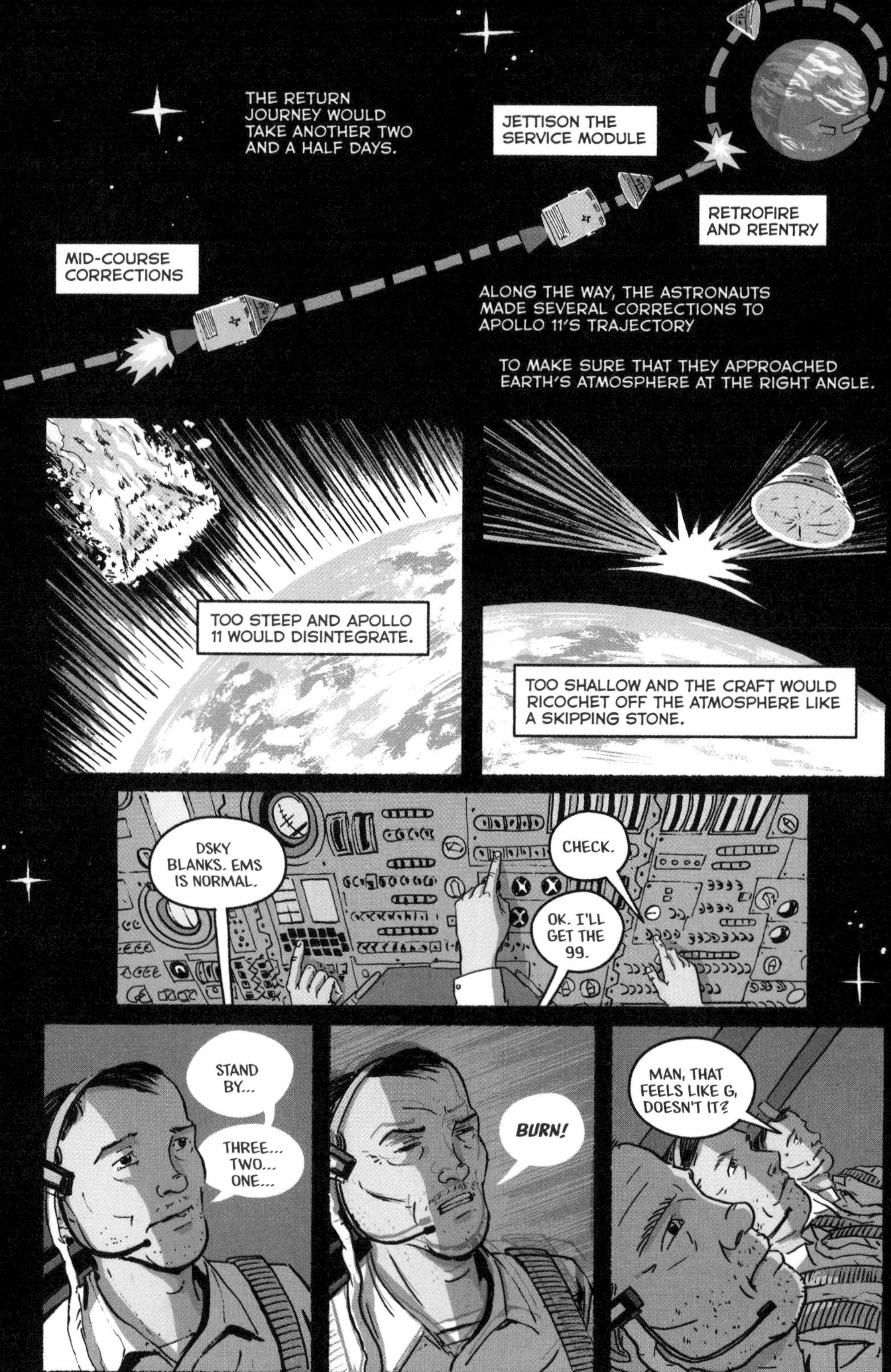
THE RETURN JOURNEY WOULD TAKE ANOTHER TWO AND A HALF DAYS.
MID-COURSE CORRECTIONS
JETTISON THE SERVICE MODULE
RETROFIRE AND REENTRY
ALONG THE WAY, THE ASTRONAUTS MADE SEVERAL CORRECTIONS TO APOLLO 11'S TRAJECTORY
TO MAKE SURE THAT THEY APPROACHED EARTH'S ATMOSPHERE AT THE RIGHT ANGLE.
TOO STEEP AND APOLLO 11 WOULD DISINTEGRATE.
TOO SHALLOW AND THE CRAFT WOULD RICOCHET OFF THE ATMOSPHERE LIKE A SKIPPING STONE.
DSKY BLANKS. EMS IS NORMAL.
CHECK.
OK. I'LL GET THE 99.
STAND BY...
THREE... TWO... ONE...
BURN!
MAN, THAT FEELS LIKE G, DOESN'T IT?

THE CREW OF APOLLO 11 SETTLED INTO A DEEP SLUMBER AS THEY SAILED EARTHWARD AT NEARLY A MILE PER SECOND.

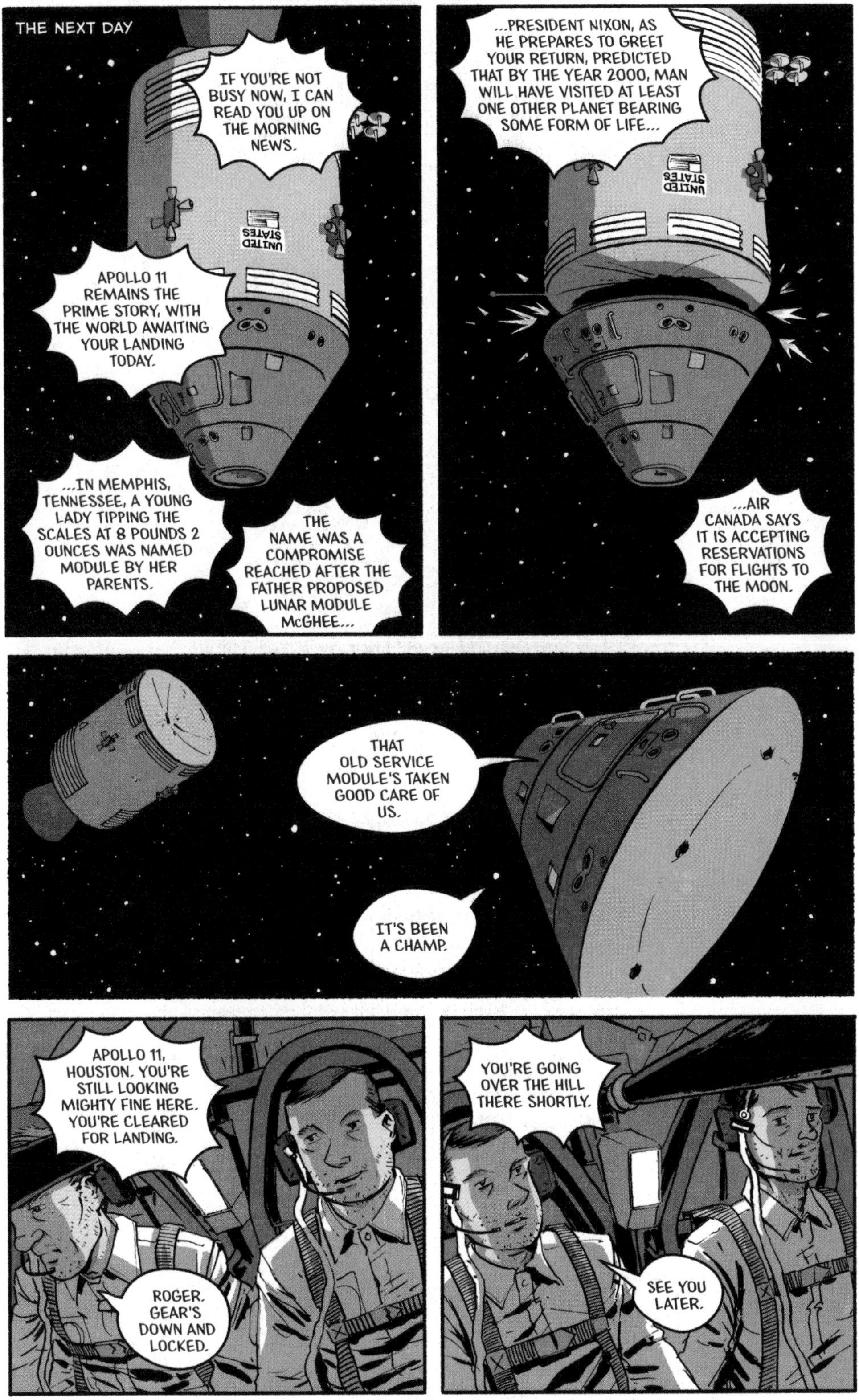
THE NEXT DAY
IF YOU'RE NOT BUSY NOW, I CAN READ YOU UP ON THE MORNING NEWS.
APOLLO 11 REMAINS THE PRIME STORY, WITH THE WORLD AWAITING YOUR LANDING TODAY.
UNITED STATES
...IN MEMPHIS, TENNESSEE, A YOUNG LADY TIPPING THE SCALES AT 8 POUNDS 2 OUNCES WAS NAMED MODULE BY HER PARENTS.
THE NAME WAS A COMPROMISE REACHED AFTER THE FATHER PROPOSED LUNAR MODULE McGHEE...
...PRESIDENT NIXON, AS HE PREPARES TO GREET YOUR RETURN, PREDICTED THAT BY THE YEAR 2000, MAN WILL HAVE VISITED AT LEAST ONE OTHER PLANET BEARING SOME FORM OF LIFE...
UNITED STATES
...AIR CANADA SAYS IT IS ACCEPTING RESERVATIONS FOR FLIGHTS TO THE MOON.
THAT OLD SERVICE MODULE'S TAKEN GOOD CARE OF US.
IT'S BEEN A CHAMP.
APOLLO 11, HOUSTON. YOU'RE STILL LOOKING MIGHTY FINE HERE. YOU'RE CLEARED FOR LANDING.
ROGER. GEAR'S DOWN AND LOCKED.
YOU'RE GOING OVER THE HILL THERE SHORTLY.
SEE YOU LATER.

ANY OBJECT--WHETHER IT'S A SPACECRAFT OR A METEOR--THAT ENTERS THE EARTH'S ATMOSPHERE AT SUCH HIGH SPEEDS

WILL BURN UP UNDER A PRIMEVAL HEAT CAUSED BY THE FRICTION OF AIR RESISTANCE.

APOLLO 11, HOWEVER, WAS PLATED WITH A SHIELD OF EPOXY AND FIBERGLASS.

STILL, THE HEAT WAS SO INTENSE THAT IT DISRUPTED ALL RADIO COMMUNICATION.

DESPITE THE TENSE SILENCE FROM APOLLO 11, THE RECOVERY CREWS CIRCLED THEIR APPOINTED PATCH OF THE PACIFIC OCEAN,

ON THE LOOKOUT FOR THE CAPSULE'S BRIGHT CANDY-STRIPED PARACHUTES.

RATHER THAN ABSORBING AND RETAINING HEAT, THIS SPECIAL ARMOR FLAKED OFF DURING REENTRY SO THAT THE HOTTEST MATERIAL WAS BLOWN AWAY FROM THE SPACECRAFT.

FROM AFAR IT LOOKED LIKE A DISASTER.

BUT INSIDE THE CAPSULE, IT WAS RELATIVELY COOL AND SAFE.

AS THE CREW WAITED FOR THE CAPSULE TO RIGHT ITSELF IN THE ROUGH SEAS, A TEAM OF NAVY SEALS APPROACHED BY HELICOPTER.

FOR THREE MEN AT THE END OF A LEGENDARY ODYSSEY, THIS WAS A MODEST HOMECOMING.

THEIR IMMEDIATE FOCUS WAS TRYING NOT TO GET SEASICK.

THE SEALS ARRIVED TO STABILIZE THE CAPSULE AND TO PASS ALONG TO THE ASTRONAUTS A SET OF SPECIAL COVERALLS.

FOR UNTIL BIOLOGISTS COULD RULE OUT THE POSSIBILITY OF LUNAR PATHOGENS, THE APOLLO 11 ASTRONAUTS HAD TO STAY IN QUARANTINE.

FROM THE CHARRED CONFINES OF THE *COLUMBIA*, THE ASTRONAUTS TRANSFERED TO THE ONLY SLIGHTLY LESS CONFINED CAPSULE OF THE MOBILE QUARANTINE FACILITY.

THE MISSION WAS OVER, BUT THE CREW OF APOLLO 11 HAD NOT QUITE REACHED THEIR JOURNEY'S END.

NEIL, BUZZ, AND MIKE. I WANT YOU TO KNOW THAT I THINK I'M THE LUCKIEST MAN IN THE WORLD, NOT ONLY BECAUSE I HAVE THE HONOR TO BE PRESIDENT OF THE UNITED STATES, BUT BECAUSE I HAVE THE PRIVILEGE OF WELCOMING YOU BACK TO EARTH.

FOR THREE WEEKS, THEY SAT IN ISOLATION FROM THE REST OF THE WORLD, COLLECTING THEIR THOUGHTS, WRITING REPORTS, AND PLAYING CARDS.

JOHN HIRASAKI--A NASA ENGINEER-- JOINED THEM IN QUARANTINE.

IT WAS HIS JOB TO UNPACK THE COMMAND MODULE AND SORT THROUGH THE LUNAR SAMPLES.

HERE JOHN, WHY DON'T YOU TAKE A LOOK?

REALLY?

IT...IT SMELLS...

...LIKE GUNPOWDER.

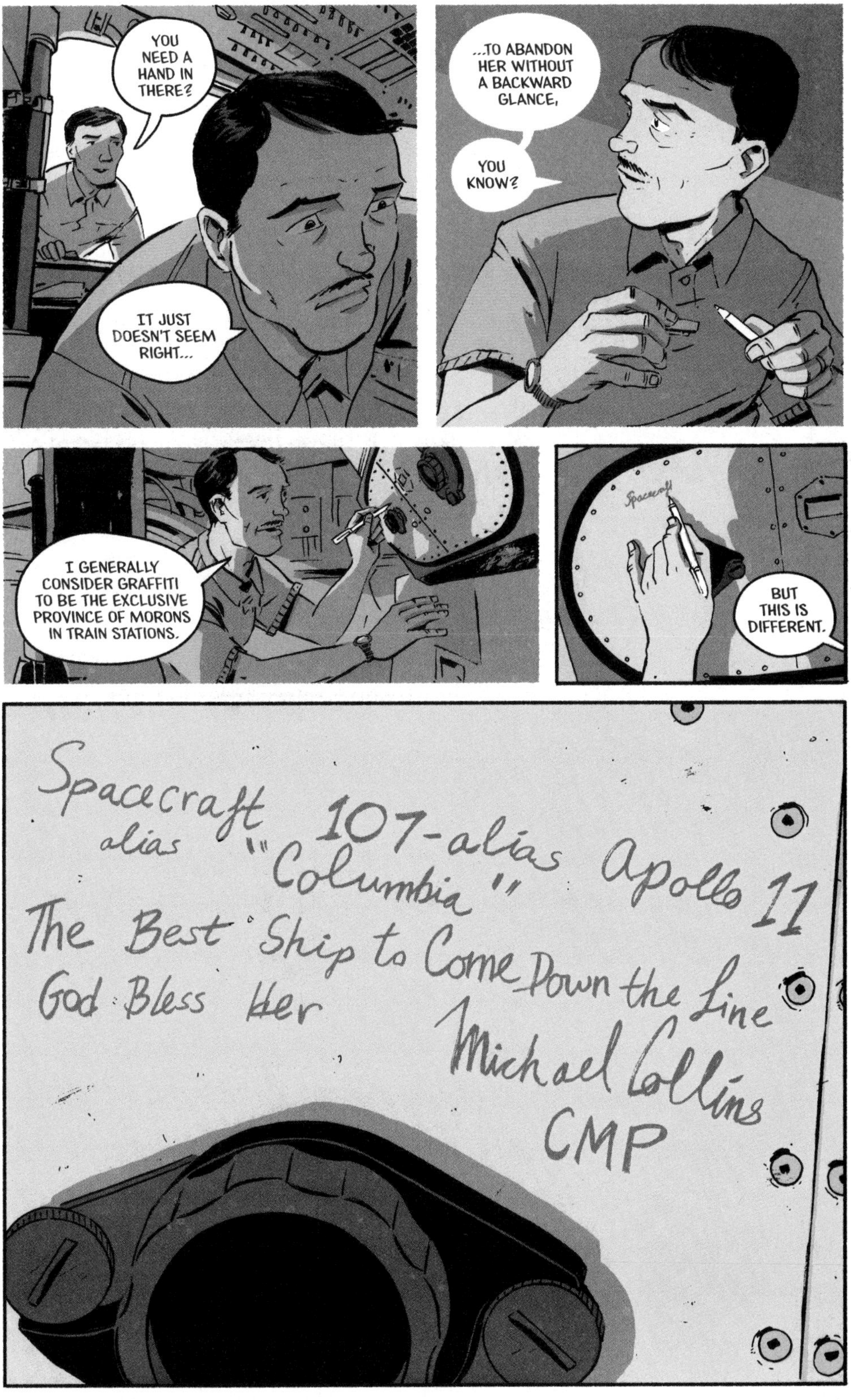
YOU NEED A HAND IN THERE?
IT JUST DOESN'T SEEM RIGHT...
...TO ABANDON HER WITHOUT A BACKWARD GLANCE,
YOU KNOW?
I GENERALLY CONSIDER GRAFFITI TO BE THE EXCLUSIVE PROVINCE OF MORONS IN TRAIN STATIONS.
Spacecraft
BUT THIS IS DIFFERENT.
Spacecraft 107-alias Apollo 11
alias "Columbia"
The Best Ship to Come Down the Line
God Bless Her
Michael Collins
CMP

EPILOGUE

SOMEWHERE OFF THE
COAST OF FLORIDA, 2012

THE RIME OF THE DEEP
OCEAN HAS A WAY OF MAKING
EVERYTHING LOOK ANCIENT.

CAN YOU
BRING US
IN A LITTLE
CLOSER?

THE TREASURE HUNTERS WHO'VE FOUND THIS
SHIPWRECK AREN'T LOOKING FOR LONG-LOST GOLD.

THE BILLIONAIRE BANKROLLING THIS
SALVAGE IS WEALTHY ENOUGH AS IT IS.

JEFF BEZOS,
THE FOUNDER
OF AMAZON

THAT LOOKS
LIKE AN
F-1 ENGINE,
RIGHT?

AND BESIDES, IT'S
TECHNICALLY STILL
NASA PROPERTY.

THIS TANGLE OF METAL IS A REMNANT OF THE FIRST STAGE OF THE APOLLO 11 ROCKET BOOSTER.

IT SAT ON THE SEAFLOOR, ABANDONED FOR FORTY-SOME YEARS.

NO ONE THOUGHT TO LOOK FOR IT,

S/N 4081951
UNIT NO 20?

AND IF ANYONE HAD, WHO WAS WILLING TO FRONT THE EXPENSE TO DREDGE IT UP?

NOW, FOUR DECADES LATER, THE CASTOFFS FROM APOLLO ARE NO LONGER JUST PIECES OF EQUIPMENT.

THEY ARE RELICS OF A BRIEF AND BYGONE AGE WHEN HUMANS FLEW TO THE MOON.

WHEN AMERICA COULD MUSTER THE MOXIE, TALENT, MONEY, AND OPTIMISM TO ACHIEVE THE IMPOSSIBLE.

OR RATHER, THAT'S ONE VERSION OF THE STORY.

IN 1970, THE STORY WAS MORE COMPLICATED.

THAT YEAR, LINDA CHARLTON, A REPORTER FOR *THE NEW YORK TIMES*, CONDUCTED AN INFORMAL POLL...

YEAH, SURE, I KNOW. LET'S SEE... IT WAS NILES SOMETHING...

MOST OF THE PEOPLE SHE SURVEYED COULDN'T REMEMBER THE NAMES OF THE APOLLO 11 ASTRONAUTS.

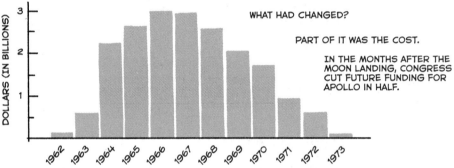

WHAT HAD CHANGED?

PART OF IT WAS THE COST.

IN THE MONTHS AFTER THE MOON LANDING, CONGRESS CUT FUTURE FUNDING FOR APOLLO IN HALF.

PART OF IT WAS THE NOVELTY. AS ONE NEWS CORRESPONDENT PUT IT:

ONCE WE LANDED ON THE MOON AND THERE WAS NOTHING THERE, IT BECAME A GEOLOGY STORY...

...AND A GEOLOGY STORY ISN'T ALL THAT SEXY COMPARED WITH THE ADVENTURE STORY OF LANDING ON THE MOON FOR THE FIRST TIME.

EVEN THE ASTRONAUTS RECOGNIZED THAT AMERICA'S ATTENTION HAD MOVED ON.

I HAD HOPED THAT THE IMPACT WOULD BE MORE FAR-REACHING THAN IT HAS BEEN...

AFTER THEIR RETURN TO EARTH, THE APOLLO 11 CREW AND THEIR FAMILIES TOURED THE WORLD IN A WHIRLWIND OF PARADES AND BANQUETS.

BUT AFTER THAT BRIEF MOMENT OF TRIUMPH CAME AN INEVITABLE LETDOWN.

I JUST CAN'T GET EXCITED ABOUT THINGS THE WAY I COULD BEFORE APOLLO 11.

BUZZ ALDRIN TOOK IT THE HARDEST.

"I HAD FLOWN TO THE MOON AND GOTTEN ITS DUST ON MY FEET.

"WHAT COULD POSSIBLY TOP THAT ACCOMPLISHMENT?

"LIFE SEEMED TO HAVE LOST ITS LUSTER.

"SOMETHING WAS WRONG; SOMETHING WITHIN ME WAS BEGINNING TO CRACK..."

HOW COULD ANYONE LIVE UP TO THE HYPERBOLE THAT THE MOON LANDING HAD INSPIRED?

RESTLESS, UNSATED BY THE UNENDING TRIALS OF HIS BRIEF EVOLUTIONARY HOUR, MAN HAS NOW MOVED FROM EARTH TO HEAVEN TO WRITE DESTINY IN BOLDER SCRIPT

IF APOLLO 11 HAD SO DEEPLY ALTERED THE COURSE OF HUMAN HISTORY, THEN WHY, BACK ON EARTH, DID THINGS SEEM TO BE GETTING WORSE?

SOLDIERS COMMITTING ATROCITIES IN VIETNAM.

CITIES COLLAPSING UNDER THE WEIGHT OF RACISM AND POVERTY.

TOXIC RIVERS CATCHING FIRE.

NONE OF THE PROBLEMS THAT SEEMED TO BE TEARING AMERICA APART IN 1969 WERE SOOTHED BY THE SUDDEN APPEARANCE OF FOOTPRINTS ON THE MOON.

ON THE EVE OF THE APOLLO 11 LAUNCH, REVERENDS RALPH ABERNATHY AND HOSEA WILLIAMS, FELLOW CIVIL RIGHTS LEADERS, MARCHED TO CAPE KENNEDY WITH HUNDREDS OF POOR AFRICAN AMERICANS.

BILLION$ FOR SPACE PENNIES FOR THE HUNGRY

END POVERTY NOW!

I AM HERE TO DEMONSTRATE WITH POOR PEOPLE IN A SYMBOLIC WAY

AGAINST THE TRAGIC AND INEXCUSABLE GULF THAT EXISTS

BETWEEN AMERICA'S TECHNOLOGICAL ABILITIES AND OUR SOCIAL INJUSTICES.

ABERNATHY WAS GIVING VOICE TO A GENERAL SENSE OF FRUSTRATION AMONG AFRICAN AMERICANS ABOUT THE MISPLACED PRIORITIES OF THE APOLLO PROGRAM.

MIXED COUPLE FIGHTS RACIST HARASSMENT

JET

EBONY

BLACK PILOT FOR WORLDS BIGGEST PASSENGER PLANE

THE MILITARY MEETS THE AFRO

...SOMEHOW WE AMERICANS MUST THINK ABOUT MAKING THE EARTH A BETTER PLACE TO LIVE. TO ESCAPE TO THE MOON IS NO ANSWER FOR ANY OF US—BLACK, WHITE, YELLOW, OR BROWN.

...ONE SMALL STEP FOR "THE MAN," AND PROBABLY A GIANT LEAP IN THE WRONG DIRECTION FOR MANKIND.

AND IT WASN'T ONLY CIVIL RIGHTS ACTIVISTS WHO QUESTIONED NASA'S PRIORITIES.

OPINION POLLS FROM THE 1960s SUGGEST THAT A MAJORITY OF AMERICANS CONSISTENTLY FELT THAT SENDING A MAN TO THE MOON SIMPLY WASN'T WORTH THE COST.

IN THE DECADES SINCE, NOSTALGIA HAS OBSCURED JUST HOW DIVISIVE THE APOLLO PROGRAM ACTUALLY WAS.

AT THE HEART OF THAT AMBIVALENCE WAS A FUNDAMENTAL CONFUSION ABOUT NASA'S MOTIVES.

EVEN PRESIDENT KENNEDY DIDN'T HAVE A CLEAR ANSWER.

WHY, SOME SAY, THE MOON?

THEY MIGHT AS WELL ASK, WHY CLIMB THE HIGHEST MOUNTAIN?

THE MOONSHOT WAS CERTAINLY MOTIVATED BY DOMESTIC POLITICS.

FOR THREE CONSECUTIVE PRESIDENTS, THE APOLLO PROGRAM WAS A USEFUL DIVERSION FROM THE GRIM REALITIES OF WAR ABROAD AND CIVIL UNREST AT HOME.

AND THE RACE TO BE THE FIRST TO LAND ON THE MOON HAD OBVIOUS STRATEGIC VALUE.

IT WAS A KIND OF PROXY WAR AGAINST THE SOVIETS, A CHANCE TO PROVE THE SUPREMACY OF THE AMERICAN WAY OF LIFE BY PLANTING A FLAG ON THE HIGHEST OF HIGH GROUNDS.

BUT WHAT PRESIDENT KENNEDY AND HIS SUCCESSORS EMPHASIZED ABOVE ALL WAS A KIND OF SPIRITUAL QUEST.

THE MOON WAS A FRONTIER, A MYTHIC PLACE BEYOND THE PALE OF CIVILIZATION, A CHANCE FOR AMERICA TO REENACT ITS ORIGIN STORY.

IN THE YEARS AFTER THE END OF THE APOLLO PROGRAM IN 1972, THE OFFICIAL JUSTIFICATIONS FOR THE MOON RACE SEEMED NAIVE AT BEST.

THE LEAKING OF THE PENTAGON PAPERS--

INTERNAL DOCUMENTS AT THE DEPARTMENT OF DEFENSE THAT REVEALED THE FUTILITY AND DISINGENUOUSNESS OF THE WAR IN VIETNAM--

AND THE IMPEACHMENT CHARGES AGAINST PRESIDENT RICHARD NIXON

I AM NOT A CROOK.

PROVOKED A GROWING DISTRUST AMONG AMERICANS TOWARD THEIR OWN GOVERNMENT.

ALL OF WHICH HELPS EXPLAIN AN UNLIKELY LEGACY OF APOLLO:

THE PERSISTENT CLAIM THAT THE WHOLE THING WAS A HOAX.

SOME FORM OF THIS CONSPIRACY THEORY HAD BEEN AROUND SINCE EVEN BEFORE APOLLO 11,

A SIGN OF ENDURING CYNICISM ABOUT THE TRUE SIGNIFICANCE OF THE MOON LANDING.

FIFTY YEARS AFTER APOLLO 11, THE MOON IS ONCE AGAIN A DISTANT DREAM.

WHILE SATELLITES SWARM IN EARTH ORBIT,

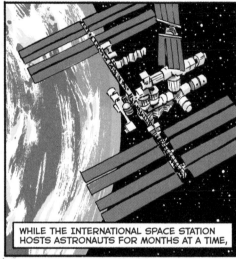

WHILE THE INTERNATIONAL SPACE STATION HOSTS ASTRONAUTS FOR MONTHS AT A TIME,

AND WHILE DEEP-SPACE PROBES ARC ACROSS THE SOLAR SYSTEM AND BEYOND,

OUR CLOSEST CELESTIAL COMPANION REMAINS OUT OF REACH.

THE LAST SATURN V ROCKET FLEW IN 1975.

WHILE THERE ARE SEVERAL MODERN ROCKETS POWERFUL ENOUGH TO REACH LUNAR ORBIT,

NOTHING YET CAN CARRY A CREW, MUCH LESS LAND THAT CREW ON THE MOON.

IN A WAY, WE'VE FOUND OURSELVES ONCE AGAIN AT THE DAWN OF A SPACE AGE,

HATCHING PLANS,

MAKING BOLD PREDICTIONS,

KINDLING A SPARK.

YET WITHOUT SOME DRAMATIC SURGE IN POLITICAL WILLPOWER, IT'S UNLIKELY THAT THERE WILL BE A NEW APOLLO PROGRAM ANY TIME SOON.

NASA HAS GRAND DESIGNS FOR AN ENORMOUS NEW ROCKET CALLED THE SPACE LAUNCH SYSTEM, OR SLS.

BIGGER THAN THE SATURN V, THE SLS WILL ALLOW NASA TO ONCE AGAIN SEND HUMANS TO THE MOON, OR EVEN MARS.

ALTHOUGH WITHOUT A NEW LANDING CRAFT, NASA WON'T BE ABLE TO ACTUALLY PUT ASTRONAUTS ON THE MOON.

MORE AMBITIOUS STILL (AND MORE SPECULATIVE) IS THE LUNAR ORBITAL PLATFORM GATEWAY,

A PERMANENT RESEARCH STATION CIRCLING THE MOON AND A POTENTIAL HARBOR FOR VOYAGES INTO DEEP SPACE.

WHILE NASA'S PLANS ARE STILL PRELIMINARY, THE SPACE AGENCIES OF OTHER NATIONS HAVE ALSO ENTERED THE RACE TO RETURN TO THE MOON.

IN JANUARY 2019, CHINA LANDED A ROBOTIC ROVER ON THE FAR SIDE OF THE MOON.

A NEW FIRST IN THE HISTORY OF LUNAR EXPLORATION.

IN THE NEXT DECADE AT LEAST SEVEN NATIONS PLAN TO SEND PROBES TO THE MOON. SOME OF THOSE COUNTRIES EVEN SPEAK OF CREWED MISSIONS.

BUT IF PLANS ARE ANY MEASURE, THE FUTURE OF LUNAR EXPLORATION BELONGS TO PRIVATE CORPORATIONS.

WHEN JEFF BEZOS DREDGED UP THE APOLLO ENGINE, IT WAS BOTH AN ACT OF PRESERVATION

AND A CANNY DISPLAY OF THE VAST PERSONAL RESOURCES AT HIS COMMAND AS THE RICHEST MAN IN THE WORLD.

SYMBOLICALLY, HOWEVER, IT WAS A DECLARATION OF INTENT, AS IF HE--AND HIS COMPANY BLUE ORIGIN--WERE PICKING UP WHERE NASA HAD LEFT OFF.

IN THIS, BEZOS IS NOT ALONE.

A HANDFUL OF HIS PEERS, WHICH IS TO SAY BILLIONAIRE ENTREPRENEURS, HAVE ALSO DECIDED THAT IT'S TIME TO RESURRECT THE LEGACY OF APOLLO.

RICHARD BRANSON WANTS HIS COMPANY, VIRGIN GALACTIC, TO OPEN UP SPACE TO TOURISM.

PAUL ALLEN'S STRATOLAUNCH AIRPLANE IS DESIGNED TO PUT SATELLITES IN ORBIT.

ELON MUSK, THE FOUNDER OF SPACEX, HOPES TO COLONIZE MARS.

THESE ARE THE FORERUNNERS IN THE NEW SPACE RACE.

2018 OFFERED A PREVIEW OF WHAT PRIVATIZED SPACEFLIGHT MIGHT LOOK LIKE.

WHEN A SPACEX ROCKET LAUNCHED A CANDY-RED CONVERTIBLE--MADE BY ELON MUSK'S ELECTRIC CAR COMPANY TESLA--INTO ORBIT,

IT WAS AT ONCE A CHEEKY STUNT, THE KIND OF IMPULSE THAT NASA WAS ALWAYS TRYING TO SUPPRESS IN ITS ASTRONAUTS--

AS WHEN ALAN SHEPARD SMUGGLED GOLF BALLS ONTO APOLLO 14--

AND AN AUDACIOUS BIT OF CORPORATE BRANDING, AN ADVERTISEMENT FOR MUSK AND HIS ASTRONOMIC AUTONOMY.

DON'T PANIC!

OF COURSE, PLANTING A FLAG ON THE MOON WAS ITS OWN KIND OF ADVERTISEMENT.

THE APOLLO PROGRAM WAS PITCHED TO THE AMERICAN PEOPLE AS A PROMISE THAT AN OPEN, DEMOCRATIC GOVERNMENT COULD RALLY ITS CITIZENS TO ACHIEVE AN UNLIKELY AND DARING DREAM.

FOR THE VANGUARD OF THE NEW SPACE AGE, THE RACE TO THE MOON IS A CONTEST FOR MARKET SHARE.

WHETHER KNOWING THIS MAKES THE NEXT FOOTPRINTS ON THE MOON FEEL LIKE ANY LESS OF A GIANT LEAP REMAINS TO BE SEEN.

AFTER ALL, HOW MANY OF HISTORY'S GREAT EXPEDITIONS WERE NOT SOMEHOW SPURRED BY THE PROMISE OF PROFIT?

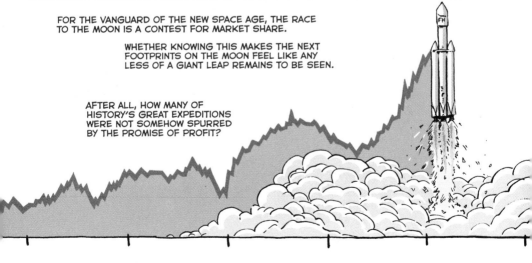

239

MEANWHILE THE MOON STILL RISES
AND SETS, AS IT ALWAYS HAS.

THE MOON STILL GROWS
AND SHRINKS BY SLIVERS,

CASTING GHOSTLY
SHADOWS IN THE NIGHT.

TWELVE MEN HAVE GONE THERE.

THEY CAME BACK WITH BOXES OF COLD,
DULL DUST FROM A DEAD WORLD.

THE MOON THAT THESE MEN WALKED UPON
IS LESS THAN OUR ANCESTORS IMAGINED.

WE CAN KNOW THAT TO BE TRUE, YET STILL
SEE THE MOON FOR WHAT IT COULD BE:

A BEACON ON THE SHORE
OF AN IMPOSSIBLY VAST
AND BARELY CHARTED SEA,

OUR CLOSEST WAYPOINT
IN ONE POSSIBLE FUTURE

WHEN WE SOMEDAY SAIL
OUT AMONG THE STARS.

241

NOTES

One unsung achievement of the Apollo Program is the sheer volume of documentation that NASA produced on its decade-long journey to the moon. These documents include mission transcripts--which I have used to make the spoken dialogue of the astronauts as accurate as possible--press materials, oral histories, and extensive photographic archives, which provided reference for many of the images in *Moonbound*. Most of the dialogue spoken by historical characters, from Kepler to von Braun and beyond, is based on written sources, which I have noted below. Any dialogue in *Moonbound* that is not explicity sourced is language that I've invented, either because of a gap in the record (often the audio from NASA missions is garbled) or for the sake of clarity.

1   **"Yes, studying the moon"**: Norman Mailer, *A Fire on the Moon* (London: Weidenfeld & Nicolson, 1970), 231.

1   **"The idea is to"**: Neil Armstrong, Michael Collins, Edwin E. Aldrin, with Gene Farmer and Dora Jane Hamblin, *First on the Moon* (New York: Little, Brown, 1970), 37.

2   **"It's nice and quiet"**: This dialogue is adapted from transcripts of audio recorded during Apollo 11, which include both the broadcast between the astronauts and Mission Control in Houston and the onboard audio captured by recorders in the actual spacecraft. For the complete transcripts (annotated with comments from flight controllers and astronauts), see David Woods, Ken MacTaggart, and Frank O'Brien, "The Apollo 11 Flight Journal," NASA, updated March 23, 2018, history.nasa.gov/afj/ap11fj/index.html, and Eric M. Jones, "Apollo 11 Lunar Surface Journal," NASA, updated December 17, 2015, history.nasa.gov/alsj/a11/a11.html.

4   **"I could, by holding"**: Armstrong et al., *First*, 19.

5   **"The lunar descent was"**: Adapted from Neil Armstrong, "Oral History of Neil A. Armstrong." Interviewed by Stephen Ambrose and Douglas Brinkley. *NASA Johnson Spacecraft Center Oral History Project*, September 2001, 86.

8   **"1202. It's fine!"**: John Garman, "Oral History of John R. Garman." Interviewed by Kevin M. Rusnak. *NASA Johnson Spacecraft Center Oral History Project*, March 2001. www.jsc.nasa.gov/history/oral _histories/GarmanJR/GarmanJR_3-27-01.htm

12   **"It has an admixture"**: Thomas L. Heath, *Greek Astronomy* (New York: Dover, 1932), 28.

16   **The ziggurats of Ur**: See Gwendolyn Leick, *A Dictionary of Ancient Near Eastern Mythology* (New York: Routledge, 1998), 152.

16   **The causeways of Tenochtitlán**: See Kay Almere Read and Jason J. Gonzalez, *Mesoamerican Mythology: A Guide to the Gods, Heroes, Rituals, and Beliefs of Mexico and Central America* (New York: Oxford University Press, 2002), 154.

16   **The Tiwi off the coast of Australia**: See Diane Johnson, *Night Skies of Aboriginal Australia: A Noctuary* (Sydney: Sydney University Press, 2014), 195.

17   **The Inuit in the far north**: See John MacDonald, *The Arctic Sky: Inuit Astronomy, Star Lore, and Legend* (Ontario: Royal Ontario Museum, 1998), 99.

17   **The Valley of the Kings**: See Robert Armour, *Gods and Myths of Ancient Egypt* (Cairo: American University in Cairo Press, 2001), 126.

17   **The Forbidden City**: See Lihui Yang, Deming An, and Jessica Anderson Turner, *Handbook of Chinese Mythology* (New York: Oxford University Press, 2008), 86.

18   **A year will have passed**: Or rather, almost a year. Twelve lunar months add up to 354 days, eleven days shorter than the sidereal year, or the time it takes the Earth to complete one orbit around the sun.

19   **"The moon is born"**: From Sermon 361 of St. Augustine of Hippo. See Bernd Brunner, *Moon: A Brief History* (New Haven: Yale University Press, 2010), 42.

19   **Were evidence of water**: As it turns out, these dark splotches are remnants of ancient lava flows. If there's any water to be found on the moon, it's probably hiding at the bottoms of deep polar craters in the form of ice.

20   **High and low tide**: There are two bulges in this diagram because the moon's gravitational pull is weaker on the far side of the Earth, weak enough that it's overpowered by the force of inertia--the tendency for moving objects (in this case the mass of water) to continue moving in a straight line. The sun's gravity also plays a part, but because it's so much farther away, it has much less influence on the Earth's tides.

21   **The year of Bur-Sagale**: W. T. Lynn, "Assyrian Eclipses," *The Observatory* 14 (August 1891), 285.

22   **Take away the sunlight**: This optical illusion is called the Gelb Effect. In 1929, the German psychologist Adhémar Maximillian Maurice Gelb observed that if you hold a black object against a dimly lit background and then illuminate it with a spotlight, the object will appear to be white. See Andrew Colman, *A Dictionary of Psychology* (New York: Oxford University Press, 2015), 305.

27   **"We can't see any"**: Because the *Eagle* landed during the lunar day, all but the brightest stars were obscured by the glare of the sun. Photographs from the mission showing a pitch-black sky are the result of the

exposure settings on the astronauts' cameras. When the astronauts themselves stood in the shadow of the LM they noticed stars that were otherwise invisible. In my drawings of lunar sky, I've split the difference.

28 **"Man hath weaved"**: John Donne, "An Anatomy of the World," *John Donne: The Major Works*, ed. John Carey (New York: Oxford University Press, 1990), 213.

30 **The breadth of his**: Kepler wrote extensively and incisively about what he perceived to be his personal shortcomings. See Arthur Koestler, *The Sleepwalkers: A History of Man's Changing Vision of the Universe* (London: Arkana, 1989), 242.

31 **And the Earth**: Apollo astronauts took many photographs of the Earth appearing to rise above the lunar horizon, but it's important to remember that all of those images were taken while in orbit. When Kepler described standing on the moon, he took care to note that from such a vantage the Earth would always be visible, day or night, and would remain essentially stationary in the sky. From the far side of the moon, of course, the Earth is never visible.

31 **"His recklessness"**: Ibid., 238.

32 **"That the sky"**: Ibid., 245.

33 **"If these do not confirm"**: Ibid., 257.

34 **"Tycho possesses"**: Ibid., 308.

35 **"Let me not seem"**: For a while there were rumors that Tycho was poisoned and that Kepler was one of the primary suspects. In 2010 a team of forensic toxicologists exhumed Tycho's remains to see if there was indeed any trace of poison, but they found no evidence of foul play.

37 **Animating this cosmic dance**: Kepler was vague about the workings of gravity, calling it simply "a force of mutual attraction." See Appendix H, "Kepler's Concept of Gravity," in Edward Rosen, *Kepler's Somnium: The Dream, or Posthumous Work on Lunar Astronomy* (Madison: University of Wisconsin Press, 1967), 219.

37 **Kepler had deciphered**: These insights have since become immortalized as Kepler's First and Second Laws of Celestial Motion, though it would take another 60 years, with the publication of the *Principia Mathematica*, before Isaac Newton formalized Kepler's theories into physical law.

38 **A German spectacle maker**: Although Lippershey was the first person to attempt to patent his device, earlier inventors either developed or at least theorized the basic principles of using lenses mounted in a tube. The conversation between Hans Lippershey and Prince Frederick Henry of Nassau is quoted in Massimo Bucciantini, Michele Camerota, and Franco Giudice, *Galileo's Telescope: A European Story*, trans. Catherine Bolton (Boston: Harvard University Press, 2015), 19.

39 **The spyglass passed**: The map showing the spread of the telescope across Europe is based on Bucciantini et al., *Galileo's Telescope*, 20–21.

40 **One evening in the fall**: Galileo was not actually the first person to observe the moon through a telescope. That honor goes to the English astronomer Thomas Harriot, who beat Galileo by four months. Harriot drew a map of lunar surface features, but his work was unpublished until 1965. See Ewen A. Whitaker, *Mapping and Naming the Moon: A History of Lunar Cartography and Nomenclature* (Cambridge: Cambridge University Press, 1999), 17–18.

40 **"The moon is by no means"**: Galileo Galilei, *Siderius Nuncius; or, the Sidereal Messenger*, trans. Albert Van Helden (Chicago: University of Chicago Press, 1989), 9–13.

42 **The moon was now another world**: As with several other innovations at the dawn of modern astronomy, the notion that the moon is a world like ours, with mountains and valleys, is much older than Galileo. The pre-Socratic Greek philosopher Democritus asserted that the moon has "lofty elevations" as well as "hollows or valleys." But then he also claimed that the Earth was a hollow disk. See Heath, *Greek Astronomy*, 38.

42 **"Four planets never seen"**: Galileo, *Siderius*, 64.

43 **"Many have since"**: This is a paraphrase of Martin Horky's attempts to discredit Galileo. For a summary of the early failures to replicate Galileo's observations, see Bucciantini et al., *Galileo's Telescope*, 90, 104.

43 **"To the noble and most excellent"**: For Kepler's letter to Galileo, see Edward Rosen, *Kepler's Conversation with Galileo's Starry Messenger* (New York: Johnson Reprint Corporation, 1965), 18.

43 **Kepler didn't yet have**: Soon he would build his own, more powerful, telescope, and he would do what Galileo never could: offer an empirical explanation for why the lenses could magnify light.

44 **"It happened one night"**: Rosen, *Kepler's Somnium*, 11.

45 **"Given ships or sails"**: Rosen, *Kepler's Conversation*, 39.

46 **As he saw it**: Quoted in Courtney Brooks, James Grimwood, and Loyd Swenson, Jr., *Chariots for Apollo: The NASA History of Manned Lunar Spacecraft to 1969* (Mineola, N.Y.: Dover, 2009), 352.

47 **He called it his Easter egg hunt**: Michael Collins describes the challenge of locating the Lunar Module in his memoir. See Michael Collins, *Carrying the Fire: An Astronaut's Journeys* (New York: Farrar, Straus and Giroux, 2019), 404.

50 **"There are many things"**: Bernard le Bovier de Fontenelle, *A Conversation on the Plurality of Worlds*, trans. Elizabeth Gunning (London: J. Cundee, 1803), 65.

51 **"Mr. Wren, what wonders"**: The lunar globe that Sir Christopher Wren presented to Charles II has since been lost, but we know about it from a brief record in the Royal Society's minutes: "This globe presented not only the spots and various degrees of whiteness upon the surface of the moon, but the hills, eminences, and cavities of it moulded in solid work." See Thomas Birch, *History of the Royal Society of London*, Vol. 1 (London: A. Millar, 1761), 21.

52 **As a result, even though**: The moon wobbles in its orbit--a phenomenon called "libration"--which means that over time, you can actually see about 59 percent of its total surface. Maybe it's worth pointing out here that the phrase "dark side of the moon" doesn't mean that the part of the moon that faces away from us is, well, dark. Any given spot on the moon experiences a cycle of one long day and one long night, each of which lasts for about 14 Earth days.

53 **A race of giants**: See Lucian, *True History*, trans. Francis Hickes (London: A. H. Bullen, 1902), 15.

53 **Snakelike creatures**: See Rosen, *Kepler's Somnium*, 28.

53 **Where murder is impossible**: See Francis Godwin, *The Strange Voyage and Adventures of Domingo Gonsales to the World in the Moon* (London: John Lever, 1768), 38.

53 **Where the old are made young**: See Cyrano de Bergerac, *A Voyage to the Moon*, trans. Archibald Lovell (New York: Doubleday and McClure, 1899), 50.

53 **Where all the lost things**: See Lodovico Ariosto, *Orlando Furioso*, trans. William Steward Rose (National Alumni, 1907), 295.

55 **"By considering the motions"**: This and the following dialogue by Newton is a paraphrase of his description of the cannon experiment. See Isaac Newton, *A Treatise of the System of the World* (London: F. Fayam, 1728), 5.

57 **"And to tell the truth"**: Jules Verne, "From the Earth to the Moon," in *Amazing Journeys: Five Visionary Classics*, trans. and ed. Frederick Paul Walter (Albany: State University of New York Press, 2010), 189.

57 **"Gallant colleagues"**: Here and in the following pages, the speech of the president of the fictitious Baltimore Gun Club appears in Verne, "Earth to the Moon," 130–141.

60 **"Look at that"**: For the scenes that take place in the Lunar Module, I have had to use my imagination to reconstruct the conversations between the astronauts. For this, I have turned to the astronauts' accounts in their memoirs as well as their recollections in the post-mission debriefing. This is a divergence from the mission transcripts. There were two ways that audio from the Apollo 11 mission made it back to Earth: radio transmissions or tapes. The tapes were recorded on devices installed in the Command Module and the Lunar Module and were recovered with the astronauts when they returned to Earth. Unfortunately, the recorder on the Lunar Module malfunctioned during the lunar descent and failed to record the onboard conversations of Armstrong and Aldrin. The thing about these onboard recordings is that they are often more candid and relaxed than the official radio transmissions (since the onboard audio wasn't being broadcast to everyone on Earth).

63 **"I am the vine"**: Aldrin's Communion ceremony was off the record because of a controversy that erupted after the crew of Apollo 8 read a passage from the Book of Genesis on a live broadcast back to Earth. Madalyn Murray O'Hair, an atheist, sued NASA, and although the case was dismissed (on grounds that lunar orbit was beyond the court's jurisdiction), NASA encouraged Aldrin to perform his solemnities in private. See Buzz Aldrin and Wayne Warga, *Return to Earth* (New York: Random House, 1973), 233.

64 **"The Earth is the cradle"**: Arkady Kosmodemyansky, *Konstantin Tsiolkovsky: 1857–1935* (Moscow: Nauka, 1985), 114.

64 **"Once the rockets"**: Tom Lehrer, "Wernher von Braun," *That Was the Year That Was* (New York: Reprise Records, 1965).

66 **"We thought nothing could hurt us"**: Adapted from the recollections of Private John Galione, 104th Infantry, as well as other members of the 104th who liberated Dora-Mittelbau. See Mary Nahas, *The Journey of Private Galione: How America Became a Superpower* (Enumclaw: Pleasant Word, 2004), 131.

68 **"I want to help turn the wheels of progress"**: See Michael J. Neufeld, *Von Braun: Dreamer of Space, Engineer of War* (New York: Vintage, 2007), 7.

68 **"It is possible to build"**: See Hermann Oberth, *Ways to Spaceflight* (NASA, 1972).

71 **That air rushes out**: This is Isaac Newton's famous Third Law of Motion: For every action there is an equal and opposite reaction.

72 **If we answer these two questions**: The Tsiolkovsky Equation, more popularly known as the "ideal rocket equation," is still used by flight planners and rocket designers. NASA astronaut Don Pettit has written a helpful introduction to the engineering challenges implicit in this equation. See www.nasa.gov /mission_pages/station/expeditions/expedition30 /tryanny.html

74 **"I was a different boy"**: For Goddard, this moment in the tree was something of an origin story. Every year he would celebrate October 19 as the "Anniversary Day" of his life's inspiration. See David A. Clary, *Rocket Man: Robert H. Goddard and the Birth of the Space Age* (New York: Hyperion, 2003), 13.

74 **"My rocket engine"**: Goddard's claims are paraphrased from his report to the Smithsonian Institution. See Robert Goddard, "A Method of Reaching Extreme Altitudes," *Smithsonian Miscellaneous Collections* 71, no. 2 (1921), 35.

75 **"To claim that it would"**: "A Severe Strain on Credulity," *The New York Times*, January 13, 1920, 12. Of course the *Times* editorial was wrong. The newspaper admitted as much 49 years later, on the day of the Apollo 11 launch: "It is now definitely established that a rocket can function in a vacuum as well as in an atmosphere. The *Times* regrets the error." See "A Correction," *The New York Times*, July 17, 1969, 43.

75 **Ever the optimist**: Clary, *Rocket Man*, 115.

76 **"My reason is"**: Letter to Smithsonian assistant secretary Charles Greeley Abbot, May 5, 1926. See Robert Goddard, *The Papers of Robert H. Goddard: 1925–1937*, 2, (New York: McGraw-Hill, 1970), 590.

78 **"My father warned me"**: Neufeld, *Von Braun.*, 97.

78 **"There can be little doubt"**: Ibid., 96.

79 **"Explain to the Führer"**: Ibid., 64.

79 **"I began to see the format of the man"**: Ibid., 110.

81 **"Gentlemen, this afternoon"**: Ibid., 137.

81 **"Lest you think"**: Michael J. Neufeld, *The Rocket and the Reich: Peenemünde and the Coming of the Ballistic Missile Era* (Washington, D.C.: Smithsonian, 1995), 165.

82 **"Pay no attention"**: Nahas, *Journey*, 103.

83 **"It is repulsive"**: Neufeld, *Von Braun*, 161.

84 **In one way**: Neufeld, *The Rocket*, 264.

85 **"We have orders"**: Neufeld, *Von Braun*, 170.

85 **"I have already"**: Ibid., 190.

86 **In the margins**: Asif Siddiqi, *Challenge to Apollo: The Soviet Union and the Space Race, 1945–1974* (Washington, D.C.: NASA, 2000), 27.

87 **"This is absolutely"**: Ibid., 24.

88 **"At forty, he is considered"**: "What Are We Waiting For?," *Collier's*, March 22, 1952, 23.

89 **"Here was a man"**: Neufeld, *Von Braun*, 461.

89 **"Within the next"**: Wernher von Braun, "Crossing the Last Frontier," *Collier's*, March 22, 1952, 25.

94 **"Whose imagination is not"**: Louis Ridenour, "The Significance of Satellite Vehicles," *Preliminary Design of an Experimental World-Circling Spaceship* (Santa Monica: Douglas Aircraft Company, Inc., 1946), 16.

95 **Whose very name**: Korolev was fanatical about the aesthetic elements of Sputnik, insisting that it be displayed on a special plinth, draped in velvet. See James Harford, *Korolev: How One Man Masterminded the Soviet Drive to Beat America to the Moon* (New York: John Wiley & Sons, 1997), 127.

97 **"When I opened my eyes"**: Korolev shared this hallucinatory memory of his last days in Siberia with cosmonauts Yuri Gagarin and Alexei Leonov just a few days before his death in 1966. Ibid., 51–53.

98 **"Our country doesn't need"**: Siddiqi, *Challenge*, 15.

99 **"Such a weapon"**: Eyewitness testimony from the American psychiatrist Robert Jay Lifton's interviews with Hiroshima survivors, quoted in Richard Rhodes, *The Making of the Atomic Bomb* (New York: Simon and Schuster, 1986), 733.

99 **"Hiroshima has shaken"**: David Holloway, *Stalin and the Bomb: The Soviet Union and Atomic Energy, 1939–1956* (New Haven: Yale University Press, 1994), 132.

100 **"We know practically"**: These are the words of Major Georgiy A. Tyulin, who was charged with producing a report on the V-2 in 1944. See Siddiqi, *Challenge*, 20.

100 **"If we increased"**: Harford, *Korolev*, 73.

101 **"Though the crystal ball"**: When this report came out, the RAND Corporation did not yet exist. The authors were members of a division of Douglas Aircraft Company dedicated to research and development. They spun off to form RAND in 1948. David Griggs, Introduction, *Preliminary Design of an Experimental World-Circling Spaceship* (Santa Monica: Douglas Aircraft Company, Inc., 1946), 1–2.

103 **"We developed a kind of"**: These are the words of Josef Gitelson. Quoted in Harford, *Korolev*, 56.

105 **"We have evidence"**: "President Truman's Statement Announcing the First Soviet Atomic Bomb, September 23rd, 1949," Atomic Archive, accessed May 6, 2018, www.atomicarchive.com/Docs/Hydrogen/SovietAB.shtml.

106 **"The successful launching"**: Paul Kecskemeti, "The Satellite Rocket Vehicle: Political and Psychological Problems," RAND RM-567, October 4, 1950, 8.

106 **"If the Soviet Union"**: A. V. Grosse, "Report on the Present Status of the Satellite Problem," August 25, 1953, ed. John Logsdon, *Exploring the Unknown: Selected Documents in the History of the U.S. Civil Space Program*, Vol. 1 (Washington, D.C.: NASA, 1995), 268.

106 **"A small scientific"**: "Draft Statement of Policy on U.S. Scientific Satellite Program," National Security Council, NSC 5520, in Logsdon, *Exploring*, 308.

106 **"By limiting the payload"**: Wernher von Braun, "A Minimum Satellite Vehicle," in Logsdon, *Exploring*, 274.

107 **"Preliminary studies"**: Logsdon, *Exploring*, 309.

108 **"The President has approved"**: Statement by James C. Hagerty, The White House, July 29, 1955, in Logsdon, *Exploring*, 200.

109 **The warhead was**: The difference between a rocket and a missile is effectively only that the latter is intended to cause damage on impact, usually by blowing up. A ballistic missile is a rocket that falls back to Earth on a ballistic trajectory, like a cannonball or bullet. In other words, any guidance or aiming of the missile happens on the way up, then gravity does the rest. Depending on the size of the engine propelling it, the ballistic missile can be anywhere from tactical (about 100 miles) to intercontinental (more than 3000 miles). The alternative is something like a Cruise missile that can be guided remotely up until the moment of impact.

109 **"The results obtained"**: From the announcement in the August 27, 1957, edition of the Soviet newspaper *Pravda*, quoted in Siddiqi, *Challenge*, 161.

112 **"Listen now for the sound"**: Yanek Mieczkowski, *Eisenhower's Sputnik Moment: The Race for Space and World Prestige* (Ithaca, N.Y.: Cornell University Press, 2013), 13.

112 **"Artificial Earth satellites"**: "Announcement of the First Satellite," *Pravda*, October 5, 1957, in Logsdon, *Exploring*, 330.

112 **"Of course it is of great"**: President Eisenhower's press secretary James Hagerty. "Mr. Hagerty's News Conference Saturday, October 5, 1957," in James Hagerty Papers, Box 49, Eisenhower Presidential Library.

112 **The American press**: These headlines appear in Paul Dickson, *Sputnik: The Shock of the Century* (New York: Walker and Company, 2001), 22–27.

113 **But according to polls**: Mieczkowski, *Sputnik Moment*, 21.

113 **"Let us not pretend"**: C. C. Furnas, "Why Did U.S. Lose the Race? Critics Speak Up," *LIFE*, October 21, 1957, 23.

113 **"Soon they will be"**: Dickson, *Sputnik*, 117.

113 **"Now, so far as"**: "The President's News Conference," October 9, 1957, Public Papers of the Presidents, item 210.

114 **The CIA had warned him**: Allen Dulles, *The Craft of Intelligence* (New York: Harper, 1963), 168.

116 **"We can put up"**: Dickson, *Sputnik*, 16.

116 **The Soviets and the Americans**: This count excludes the six failed tests of Project Pilot, the U.S. Navy's attempt at an air-launched orbital vehicle.

120 **"I know from preflight"**: All the text in this chapter is from Collins, *Carrying the Fire*, 402.

124 **"A man should have"**: Tom Wolfe, *The Right Stuff* (New York: Farrar, Straus and Giroux, 1983), 24.

126 **Dr. Lovelace pioneered**: Actually, the true pioneer of "space medicine" was the Nazi physician Dr. Hubertus Strughold, whose notorious experiments on prisoners at Dachau concentration camp did not disqualify him from American citizenship after the war as part of Operation Paperclip.

129 **Psychiatrists probed**: "Sample Questions from Project Mercury Tests," NASA, accessed May 8, 2018, history.nasa.gov/40thmerc7/samplequestions.pdf.

129 **For that, they would need**: See Roger D. Launius, "Heroes in a Vacuum: The Apollo Astronaut as Cultural Icon," *The Florida Historical Quarterly* 87, no. 2, 174–209.

130 **"What do your wives"**: These questions and answers are adapted from the transcript of the press conference announcing the Mercury 7 crew, April 9, 1959. See "Press Conference of Mercury Astronaut Team," NASA, accessed May 8, 2018, history.nasa.gov/40thmerc7/presscon.pdf.

131 **"They're applauding us"**: Alan Shepard and Donald Slayton, with Jay Barbree and Howard Benedict, *Moonshot: The Inside Story of America's Race to the Moon* (Atlanta: Turner Publishing, 1994), 65.

131 **Demure, supportive wife**: The Astronaut Wives, as they were labeled in the press, bristled at these simplifications, while the astronauts themselves were sometimes less than ideal husbands. See Lily Koppel, *The Astronauts' Wives Club* (New York: Grand Central Publishing, 2014), 36.

131 **The astronaut Everyman**: Add to this list an implicit qualifier: white-skinned. The first black test pilot to apply to become an astronaut was Edward Dwight, Jr., in 1963. He and 300 other white candidates didn't make the cut. In 1967 Robert Lawrence, Jr., became the first African-American astronaut as part of an Air Force reconaissance program. He died in a plane crash six months later, and the Air Force program was canceled before it flew any missions. See John Charles, "A Hidden Figure in Plain Sight," *The Space Review*, June 12, 2017, www.thespacereview.com/article/3262/1.

132 **But was the work**: In 1960 Betty Skelton, a pilot and aerobatics champion, was the subject of a photo essay in *Look* magazine. NASA let her train alongside the Mercury astronauts as a kind of publicity stunt. Although Skelton passed the tests, she was never seriously considered a candidate for spaceflight. She

recalled, "I felt it was an opportunity to try to convince them that a woman could do this type of thing and could do it well." See Margaret Weitekamp, *Right Stuff, Wrong Sex: America's First Women in Space Program* (Baltimore: Johns Hopkins University Press, 2004), 65–69, and Ben Kocivar, "The Lady Wants to Orbit," *Look*, February 2, 1960, 112–119.

132 **"Certain qualities"**: Weitekamp, *Right Stuff*, 77.

133 **"Miss Cobb is qualified"**: Quoted in "From Aviatrix to Astronautrix," *Time*, August 19, 1960.

134 **When Lovelace announced**: Ibid. For other reactions to Cobb's tests, see Weitekamp, *Right Stuff*, 78.

135 **"Regret to advise"**: Ibid., 117.

135 **"Let's stop this"**: Ibid., 137.

135 **"The men go off"**: John Glenn's testimony, in "Qualifications for Astronauts: Hearings Before the Special Subcommittee on the Selection of Astronauts of the Committee on Science and Astronauts, U.S. House of Representatives, Eighty-seventh Congress, Second Session, July 17 and 18, 1962," Vol. 2, 67.

135 **"Only if NASA gives us"**: This was a common joke inside NASA at the time. Von Braun is quoted in Howard E. McCurdy, *Space and the American Imagination* (Baltimore: Johns Hopkins University Press, 2011), 295.

135 **The United States lost**: While Tereshkova was the first woman in space, she was also an outlier. Another 19 years would pass before the flight of the second female cosmonaut, Svetlana Savitskaya, in 1982, followed a year later by the first American female astronaut, Sally Ride. In that span of time, male astronauts flew to space 166 times.

137 **"I want you guys"**: Hamish Lindsay, *Tracking Apollo to the Moon* (London: Springer, 2001), 44.

137 **"Poyekhali"**: Yuri Gagarin's famous exclamation during the launch of his Vostok 1 mission loosely translates to "Let's go!"

138 **"I see the Earth"**: Paraphrased from the Vostok 1 mission transcript. See Siddiqi, *Challenge*, 277.

138 **Or NASA's spokesman**: Jay Barbree, *"Live from Cape Canaveral": Covering the Space Race, from Sputnik to Today* (New York: HarperCollins, 2007), 52.

139 **"Is there any other"**: John F. Kennedy, "Memorandum for Vice President," April 20, 1961, in John Logsdon, *Exploring the Unknown: Selected Documents in the History of the U.S. Civil Space Program*, Vol. VII (Washington, D.C.: NASA, 2008), 479.

140 **"The exploration of space"**: President Kennedy's speech at Rice University was a follow-up to an earlier address that the president had delivered to Congress, requesting funding for the Apollo Program. For the text of the Rice speech, see "John F. Kennedy Moon Speech--Rice Stadium, September 12, 1962," NASA, accessed May 8, 2018, er.jsc.nasa.gov/seh/ricetalk .htm.

140 **"You have an idiopathic"**: Donald K. Slayton with Michael Cassutt, *Deke!: U.S. Manned Space: From Mercury to the Shuttle* (New York: Tom Doherty Associates, 1994), 115.

141 **We were going to the moon**: The silver lining is that Deke made some major lifestyle changes and was eventually cleared for spaceflight on the Apollo-Suyez Test Project in 1975. He spent 217 hours in orbit. Slayton, *Deke!*, 282.

142 **Kennedy's call**: Dollar amounts adjusted for 2018 inflation. For a breakdown of NASA's funding during the Apollo Program, see Brooks et al., *Chariots for Apollo*, 409.

142 **And, of course**: Including Neil Armstrong, who became an astronaut in the next round of selections in 1962, followed by Aldrin and Collins, who joined NASA as members of Astronaut Group 3 in 1963.

148 **"When you consider"**: Quoted in David Mindell, *Digital Apollo: Human and Machine in Spaceflight* (Cambridge, Mass.: MIT Press, 2011), 66.

148 **"Control of space means"**: "Statement of Democratic Leader Lyndon B. Johnson to the Meeting of the Democratic Conference on 7 January 1958," Statements of LBJ Collection, Box 23, Lyndon Baines Johnson Library, Austin, Texas.

149 **No, honey**: This scene is adapted from an anecdote in Hamilton's keynote address delivered to the International Conference on Software Engineering. See Margaret Hamilton, "The Language as a Software Engineer" (speech, Gothenburg, Sweden, May 31, 2018), ICSE, www.icse2018.org/info/keynotes.

151 **That was code**: This is Janez Lawson. Trained as a chemical engineer (but unable to find work in the field because of her gender and skin color), Lawson was hired as a computer at the Jet Propulsion Laboratory in Pasadena, California. She was the first African-American at the lab to hold a technical position. See *Rise of the Rocket Girls: The Women Who Propelled Us, from Missiles to the Moon to Mars* (New York: Little, Brown, 2016), 81.

151 **"Engineers make up"**: Quoted in Holt, *Rise of the Rocket Girls*, 166.

152 **Even further from**: See Margot Lee Shetterly, *Hidden Figures: The American Dream and the Untold Story of the Black Women Mathematicians Who Helped Win the Space Race* (New York: William Morrow, 2016).

152 **Second, NASA itself**: These statistics come from Sylvia Doughty Fries, *NASA Engineers and the Age of Apollo* (Washington, D.C.: NASA, 1992), 205.

152 **In the words of**: Kim McQuaid, "'Racism, Sexism, and Space Ventures': Civil Rights at NASA in the Nixon Era and Beyond," in *Societal Impact of Spaceflight*, ed. Stephen Dick and Roger Launius (Washington, D.C.: Government Printing Office, 2007), 424.

153 **"I changed what I"**: Quoted in Shetterly, *Hidden Figures*, 263.

154 **"Astronauts don't make"**: As it turned out, this exact error occurred during Apollo 8 after astronaut Jim Lovell entered the wrong command into the computer. Hamilton and her team were on call during the mission and jury-rigged a fix to reboot the computer. See "Margaret Hamilton's Introduction," July 27, 2001, Apollo Guidance Computer History Project, authors.library.caltech.edu/5456/1/hrst.mit.edu/hrs /apollo/public/conference1/hamilton-intro.htm.

155 **"We like to think"**: Quoted in Mindell, *Digital Apollo*, 67.

156 **Just to the east**: Korolev died two weeks before Luna 9 landed on the moon. His health had deteriorated in the early 1960s, and the intense pressure from Soviet leadership to constantly outdo the American space program probably had something to do with it. In his obituary, the name of the Chief Designer, a state secret for over a decade, was finally made public.

157 **"It is man"**: Quoted in John Logsdon, *The Decision to Go to the Moon* (Chicago: University of Chicago Press, 1976), 125.

157 **"I'm not that interested"**: See "A Historic Meeting at the White House on Human Spaceflight," NASA History, updated November 5, 2002, history.nasa .gov/JFK-Webbconv/index.html.

158 **"Now we can get rid"**: Quoted in an editorial, "Let Man Take Over," in *The New York Times*, February 25, 1962, E 10.

159 **"I'm separating from"**: Edward White and James McDivitt, Gemini IV.

159 **"Fourteen days"**: James Lovell, Jr., Gemini VII.

159 **"I've been struggling"**: James McDivitt, Gemini IV.

159 **"We're all sitting"**: Walter Schirra, Jr., Gemini VI-A.

162 **"We can detect"**: Paraphrased from Gene Kranz, *Failure Is Not an Option: Mission Control from Mercury to Apollo 13 and Beyond* (New York: Simon and Schuster, 2000), 253.

162 **"If you viewed it"**: Quoted in Rick Houston and Milt Heflin, *Go, Flight! The Unsung Heroes of Mission Control, 1965–1992* (Lincoln: University of Nebraska Press, 2015), 56.

162 **"It's an IBM 7094"**: Paraphrased from interview, see NASA Johnson Space Center Oral History Project, March 10, 2010, www.jsc.nasa.gov/history /oral_histories/GarciaH/GarciaH_3-10-10.htm.

162 **"We would get a computer"**: Catherine T. Osgood, interview, NASA Johnson Space Center Oral History Project, November 15, 1999, www.jsc.nasa.gov /history/oral_histories/OsgoodCT /OsgoodCT_11-15-99.htm.

170 **Dee O'Hara gave herself**: Dee O'Hara recalls her morning in Armstrong et al., *First*, 35.

175 **Near the launch site**: Although the VAB was superseded in 1967 by the reigning champion of voluminous buildings--the Boeing airplane factory in Everett, Washington--it still holds the record for the world's largest single-story building.

177 **"You're it, guys"**: Shepard and Slayton, *Moonshot*, 237.

178 **"Well, I didn't have"**: Armstrong et al., *First*, 44.

178 **"They don't pay me"**: Ibid., 44.

179 **"We got a bad fire"**: The garbled recordings of that last, desperate cry from Apollo 1 make it difficult to say for sure what was said. This is one possibility.

180 **"I'd better not describe"**: For accounts of the pad fire, see Barbree, *Live*, 128, and Andrew Chaikin, *A Man on the Moon: The Voyages of the Apollo Astronauts* (New York: Penguin, 1994), 12.

181 **"If we die"**: Brooks et al., *Chariots for Apollo*, 220.

184 **Early prototypes**: These "hard" suits were mechanically elegant and they looked impressively space-aged, but they were far more cumbersome than the "soft" suits devised by ILC. See Nicholas de Monchaux, *Spacesuit: Fashioning Apollo* (Cambridge, Mass.: MIT Press, 2011), 240.

189 **The lunar morning**: NASA figured that the ideal angle of the sun was between 7 and 20 degrees above the horizon, which, given that a day on the moon lasts the equivalent of 14 Earth days, is why the right lighting conditions for any given location happen only once every lunar cycle.

190 **Second, if it's at**: Landing on the far side of the moon was a nonstarter, since that would mean the astronauts would be out of radio contact with Mission Control.

190 **Working backward**: Like everything having to do with spaceflight, there's way more to it than that, including all the calculations involved in deciding at what hour a launch has to happen in order to send astronauts on a trajectory to intercept the moon, a trajectory aimed not at where the moon is, but where it *will* be.

190 **"We could train"**: Chaikin, *Man on the Moon*, 177.

191 **"I waited alone"**: Aldrin and Warga, *Return to Earth*, 218.

191 **"That wondrous white machine"**: Jennifer Bogo, "The Oral History of Apollo 11," *Popular Mechanics*, July 17, 2018, www.popularmechanics.com/space /moon-mars/a4248/oral-history-apollo-11/.

191 **"It's a key to the moon"**: Armstrong et al., *First*, 65.

194 **"All engine running"**: Jack King, the Public Affairs Officer on Apollo 11, was an otherwise unflappable voice during the launch countdowns, but in this moment his emotions got the better of him—he clearly meant to say "engines," plural.

198 **One of NASA's less**: Sometimes called NASA-speak or NASA-ese. For a helpful glossary, see Paul Dickson, *A Dictionary of the Space Age* (Baltimore: Johns Hopkins University Press, 2010).

198 **"A beautiful sight!"**: John Glenn, Mercury VI.

198 **"It's been a beautiful"**: Gordon Cooper, Mercury IX.

198 **"What a view"**: Gus Grissom, Gemini III.

198 **"Sure is beautiful"**: Jim McDivitt, Gemini IV.

198 **"Of course I thought"**: Armstrong et al., *First*, 296.

198 **"We weren't trained"**: Collins, *Carrying the Fire*, 54. For a cheeky take on tongue-tied astronauts, see William H. Honan, "Le Mot Juste for the Moon," *Esquire*, July 1969, 53.

199 **The goal of those early**: This dialogue is from the mission transcripts of Apollo 7, 8, 9, and 10.

200 **Foil-thin walls**: According to Michael Collins's memoir, on more than one occasion someone dropped a screwdriver while working on the Lunar Module, and the tool actually fell through the wall of the spacecraft. See Collins, *Carrying the Fire*, 324.

201 **To give the audience**: The term "zero G" does not actually refer to the absence of gravity, but rather the absence of G-force, the acceleration that causes the feeling of weight.

205 **Collins thought**: Collins, *Carrying the Fire*, 387.

207 **"That's one small step"**: There have been several attempts to pin down exactly what Armstrong said when he first stepped onto the lunar surface. He contends that the phrase he spoke was "That's one small step for *a* man, one giant leap for mankind," which makes the most sense grammatically. But in recordings it sounds like Armstrong dropped the "a." Researchers at Michigan State University make a compelling case that English speakers from central Ohio (e.g., Armstrong) tend to pronounce the words "for a" as one word that sounds like "furrah." See "MSU-Led Team Deciphers Famous Moon-Landing Quote," *MSUToday*, June 3, 2013, msutoday.msu.edu /news/2013/msu-led-team-deciphers -famous-moon-landing-quote/.

211 **"Hello, Neil and Buzz"**: President Nixon's speechwriter, William Safire, had also drafted a somber message for the president to deliver on the chance that the Apollo 11 landing failed. It starts: "Fate has ordained that the men who went to the moon to explore in peace will stay on the moon to rest in peace." See "In Event of Moon Disaster," National Archives, updated August 15, 2016, www.archives.gov/presidential-libraries/events /centennials/nixon/exhibit/nixon-online-exhibit -disaster.html.

212 **"Houston, Columbia"**: One of the more stubborn misconceptions used to bolster arguments that the moon landing was a hoax is the way that, in photographs from the moonwalk, the flag seems to flap in the wind. There's no wind on the moon because there's no atmosphere. The flag was simply wrinkled from being folded up during the flight.

214 **Later, Aldrin would reflect**: *Apollo 11 Technical Crew Debriefing* (Houston: NASA Manned Spacecraft Center, 1969), Vol. 1, Sec. 10, 33.

214 **"Adios, amigo"**: Let the record books state that, for now, there have been *two* languages spoken on the moon.

226 **"Neil, Buzz, and Mike"**: Armstrong et al., *First*, 362.

226 **"It..it smells..."**: Paraphrased from an interview with John K. Hirasaki. See "NASA Johnson Space Center Oral History Project," NASA, March 6, 2009, www.jsc .nasa.gov/history/oral_histories/HirasakiJK /HirasakiJK_3-6-09.ht m.

227 **"It just doesn't seem"**: Collins, *Carrying the Fire*, 446.

230 **Most of the people**: Linda Charlton, "Check Finds Many Forget Apollo 11," *The New York Times*, July 19, 1970, 54.

230 **"Once we landed"**: Reuters correspondent Mark Bloom, quoted in David Meerman Scott and Richard Jurek, *Marketing the Moon: The Selling of the Apollo Lunar Program* (Boston: MIT Press, 2014), 122.

230 **"I had hoped"**: Neil Armstrong is quoted in Matthew Tribbe, *No Requiem for the Space Age: The Apollo Moon Landings and American Culture* (New York: Oxford University Press, 2014), 9.

231 **"I just can't get"**: Collins, *Carrying the Fire*, 462.

231 **"I had flown to the moon"**: Buzz Aldrin and Ken Abraham, *Magnificent Desolation: A Long Journey Home from the Moon* (New York: Harmony, 2009), 80.

232 **"Restless, unsated"**: Israel Shenker, "Throughout History, Restless Men Have Always Been Lured by the Unknown," *The New York Times*, July 21, 1969, 11.

233 **"I am here to demonstrate"**: Reverend Ralph Abernathy, quoted in Simeon Booker, "Blacks Scarce as Men on Moon at Launch," *Jet*, July 31, 1969, 9.

233 **"Somehow we Americans"**: Simeon Booker, "Moon Probe Laudible--But Blacks Need Help," *Jet*, July 31, 1969, 10.

233 **"...One small step"**: "How Blacks View Mankind's 'Giant Step,'" *Ebony*, September 1970, 33.

233 **Opinion polls**: The most dramatic exception to this trend was during the Apollo 11 flight, when polls showed that more the 50 percent of Americans surveyed expressed support for the moon landing. See Roger Launius, "Public Opinion Polls and Perceptions of US Human Spaceflight," *Space Policy* 19 (2003), 163.

235 **The persistent claim**: If you've made it this far, you probably don't need to be further convinced that the moon landing did in fact occur, but nevertheless, for a thorough debunking, see Robert Myers and Robert Pearlman, "Apollo Moon Landing Hoax Theories That Won't Die," *Space.com*, September 2, 2011, www .space.com/12814-top-10-apollo-moon-landing-hoax -theories.html.

235 **Some form of this conspiracy**: Andrew Chaikin, "Live from the Moon: The Societal Impact of Apollo," Dick and Launius, *Societal Impact*, 63.

ACKNOWLEDGMENTS

Thank you to Katie Zanecchia and Amanda Moon for pointing me in the right direction at the outset of this project, and to Howard Yoon and Laird Gallagher for shepherding *Moonbound* into print. Mike McWhinnie was an invaluable reader, given his deep familiarity with both NASA history and orbital mechanics. Sinda Puryer is a master librarian and virtuoso of interlibrary loans. Scott Mahr lent me his engineering acumen and attention to detail. Greg Wayne was, as always, a patient tutor in all things having to do with numbers. Gregor Sokol advised me on the brief bits of dialogue that appear in Polish. Consultations with George and Elma Giavasis and Tom Biby helped transform half-finished sketches and half-baked ideas into full-fledged artwork. Laura Martin then gave much of that artwork a splash of color. The lion's share of my gratitude, however, belongs to Charlotte Housel, without whose insight, patience, and laughter this book would never have been.